AF406282

I Am; Therefore I Think

Consciousness and Humanity in the Age of AI

JP Pulcini

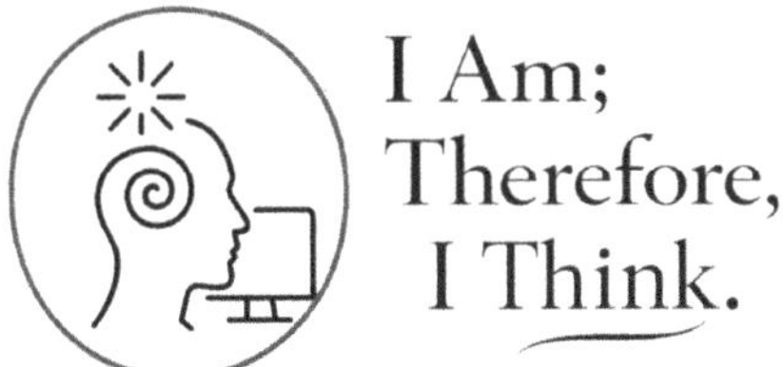

*For my family who provides
purpose and meaning in my life*

TABLE OF CONTENTS

PREFACE

The subject of consciousness has been on my mind of late—no pun intended. Maybe that curiosity began long before I realized it. Growing up in an Italian Catholic home in Boston, I was taught about the Holy Spirit, Jesus, God, saints, miracles, and Heaven and Hell as if they were part of the everyday world. At five years old, the universe felt ordered and secure. Things made sense simply because they were supposed to.

But the older I became—and the more the world changed around me—that sense of certainty began to shift. Advancements in technology, space exploration, and medicine challenged the boundaries I thought were fixed. Each breakthrough raised new questions about the soul, consciousness, and what it means to be human. The world no longer felt as neatly defined as it once did.

Somewhere along the line, I started wondering about things I never thought I'd question:

Can we ever really understand what we are?

Could a machine ever approximate a soul?

And what happens when technology encroaches on the spaces where we once placed mystery, spirit, and meaning?

Those questions resurfaced with force when I began studying AI for my MBA after thirty years in the IT field. At first, it was a practical step—a way to stay relevant in a profession that many now fear will be reshaped or replaced by artificial intelligence. But as I dug deeper into AI, something unexpected happened: the conversation expanded. It wasn't just about staying employable anymore. It became a doorway back to existential questions I hadn't asked since sitting in philosophy courses reading Aristotle, Aquinas, Socrates, and, of course, Descartes and his famous line, *"I think, therefore I am."*

What Descartes wrote about thinking and being made sense in the 17th century. In today's world—with machines that can mimic reasoning, generate language, and even simulate emotion—that statement hits differently.

Pop culture has been wrestling with these questions, too. Movies like *The Matrix* and *Blade Runner* use fiction to ask what philosophers have debated for centuries:

What is real?

Where does identity come from?

What distinguishes a human being from something that looks, thinks, or remembers like one?

Those films exaggerate, yes, but they point to something essential: once we start questioning the nature of our own experience, we're already knee-deep in the mystery of consciousness. And doubt, I've realized, is not a flaw—it's evidence. If we can question our world, our memories, our purpose, or whether any of this is "real," then something undeniably human is doing the questioning.

But doubt gets messier in a world shaped by AI. Machines can now stretch into territories once reserved for the mind: decision-making, pattern recognition, creativity, and maybe someday, self-modeling. Technology lengthens our lives, alters our bodies, and reshapes how we interact with the world. The boundaries between mind, body, and machine begin to blur—sometimes subtly, sometimes dramatically.

And that is where this book begins.

I'm not an academic or a philosopher by trade. I'm a lifelong technologist, a reader, and someone who's always tried to make sense of how things work—both the things we build and the things we carry inside us. This book isn't an attempt to prove anything or defend a theory. It's an exploration: part personal, part philosophical, part cultural, and part speculative.

It's a journey through the convergence of mind, body, and technology—and an attempt to understand what these shifts mean for all of us as we move into a future where the definition of "human" may not be as clear as it once was.

So, I invite you to walk through these questions with me.
Not for answers—because I don't have them—but for a deeper understanding of what it means to think, to wonder, and to be conscious at a time when consciousness itself is up for debate.

JP

PART I—BEING HUMAN

"As I observed that this truth, I think, therefore I am (COGITO ERGO SUM), was so certain and of such evidence that no ground of doubt, however extravagant, could be alleged by the sceptics capable of shaking it, I concluded that I might, without scruple, accept it as the first principle of the philosophy of which I was in search."

Discourse on Method, Part IV—René Descartes

CHAPTER 1

MY MIND WANDERS AND WONDERS

THE LIGHT IS BRIGHT, AND THERE IS A SENSATION OF warmth. A feeling of peace and calm comes over me, but I don't know why. I know it feels good, and at least in that moment, I know I'm here now, and I'm alive. This is one of my earliest memories, *I think.*

Consciousness begins long before we learn the word for it. Before philosophy, before science, and before any attempt to describe the mind, each of us encounters something unmistakably real: *the quiet awareness that we are here.* A flicker of memory, a moment of wonder, and a feeling we cannot yet name—these early impressions become the first signs of a mind noticing itself.

Every human life starts this way, not with definitions or beliefs, but with experience: warmth, sound, sensation, safety, and fear. Somewhere within that early blur, awareness takes shape. We sense that the world is larger than the moment we are in. Something in us begins to reach outward and inward at the same time, asking not simply *What is this? But what does this mean for me? And who might I become inside a world this vast?*

This book begins in that space—the origin of consciousness and the unfolding of the wandering mind. For as long as humans have existed, we have tried to understand the spark inside us: the driving force that makes us curious, reflective, imaginative, restless, hopeful, and aware. We think, yes. But we also dream, doubt, create, question, remember, and reach for meaning beyond what our senses provide us.

Today, as artificial intelligence (**AI**) grows ever more capable and more present in our lives, the question becomes even more urgent: *What exactly is this thing we call consciousness?* And why does it feel so different from anything that we can build in a machine?

To explore this mystery, we begin with the earliest layers of the human mind—memory, sensation, and wonder—the building blocks that allow consciousness to form and for identity to take root. It is from these first sparks of awareness that our entire inner world eventually unfolds, and where we start this journey.

• •

Section 1—Early Human Memory and Wonder

CONSIDER YOUR FIRST MEMORIES, OR EVEN THE EARLIEST memories any human being may have recorded once language emerged and writing became possible. Were those memories pleasant or painful, traumatic or euphoric, or something in between? More importantly, how were these earliest experiences preserved before language or written symbols existed? Most likely, they get passed on through stories told around campfires—oral histories carried across generations until writing and structured language became common enough to capture them.

The Paleolithic cave paintings at Lascaux, France, date back between 17,000 and 22,000 years. Yet *Homo sapiens* themselves extend back at least 300,000 years in Africa (2017 dating of Jebel Irhoud fossils, now ~315k years old). Evidence of more culturally "modern" behavior—art, symbolic expression, and sophisticated tools—appeared around 40,000-50,000 years ago. Regardless of where one stands on human origins, our progression from small bands of hunter-gatherers to complex civilizations reflects something remarkable*: a unique capacity for abstraction, reasoning, self-reflection, and symbolic thought.* That capacity—our ability to think—has long been seen as the defining trait that separates humans from other animals.

For many, theology intertwines with this belief: the idea that human beings are endowed by a divine creator (or creators) with a special gift—reason, free will, and a mind capable of contemplating itself. Whether taken literally or metaphorically, this ability to think, to reason, to doubt, and to question everything is at the core of how we discover meaning and evolve as a species. It is the spark that allows us to imagine the future, reinterpret the past, communicate experiences, express new ideas, and continually reshape our understanding of the world.

When Descartes proclaimed *"I think, therefore I am,"* he was articulating a form of *dualism*—arguing that mind and body are distinct substances that interact much as a passenger interacts with a vehicle. In this Cartesian view, thought itself is the primary evidence of existence. Consciousness—our ability to think, reflect, or doubt—is the undeniable proof that we are more than mere physical bodies responding to stimuli. It suggests that something about the mental world is irreducible and independent, serving as the seat of identity, free will, and awareness.

This shift from *thinking* to *wondering* is where I want to begin. Most people hold early memories from age three to four: a comforting gesture from a parent, a taste of something sweet, or the sensation of a warm breeze across the face. These simple sensory impressions become the first anchors of experience. They also mark the beginning of wonder—our innate desire to ask why, to explore, and to find meaning in what we perceive. Wonder begins the moment that we realize the world is bigger than our immediate sensations.

••

Section 2—The Island Thought Experiment

IMAGINE, FOR A MOMENT, A PERSON BORN ON A BEAUTIFUL BUT deserted island, raised by a single caretaker who learned from the person before them. Their world would be shaped entirely by what the island provides—food, weather, terrain, tools they can fashion, and whatever survival skills they inherit. Like Tom Hanks' character in *Cast Away*, they might learn to fish, build shelter, make fire, and even craft rudimentary clothing. But everything beyond the island—civilization, language, culture, and broader society—would be unknown, limited to a small shared history told by their caregiver or what they carried when they were deserted on that island.

Yet at some point, this person would look toward the ocean and wonder: *What lies beyond the horizon?* With time, curiosity might drive them to build a boat, explore, and push beyond the boundaries of their known world. This shift—from simple thinking to active wondering—is fundamental. It reflects a uniquely human impulse: the drive to seek, learn, understand our existence, and pursue meaning beyond what we can immediately see or know.

It is from this starting point—this interplay between memory, experience, thought, and wonder—that we begin our exploration of consciousness. And it raises the central question for this chapter: *Is consciousness tied to the curiosity that compels us to look beyond our immediate world?*

Perhaps consciousness is not just awareness, but the desire to push past the edges of the known—to search for meaning, to question our place in the universe, and to wonder what exists beyond the horizon of our current understanding. That wondering is what drives exploration outward—from our inner desire to always seek meaning in the world.

• •

Section 3—Wonder, Mind, and First Memory

As we consider 'wonder' not merely as a reaction but as a cognitive act, it becomes clear that wondering is fundamentally different from simple thinking. Thinking can be mechanical—processing, recalling, calculating, and reacting. Wondering, however, requires the mind to *step outside* of its immediate experience. When you wonder, you project yourself into a space that does not yet exist: a possibility, an unknown, an imagined future, or a question that has no answer yet. That mental leap—beyond the constraints of the present moment—is one of the earliest and most profound indicators of consciousness.

Philosophers often describe consciousness as the "subjective experience of being"—Thomas Nagel's famous "what it is like" to be something. But before we ever encounter philosophical language, each of us experiences this concept firsthand in childhood. Our earliest memories are rarely intellectual; they are sensory, emotional, and fleeting. Yet they become the first building blocks of our internal world. They shape who we believe ourselves to be and what we can imagine becoming. They are the earliest evidence that the mind does more than observe the world —it *interprets* it.

I remember one of my own earliest memories with surprising clarity—not because it was highly dramatic or visual, but because of the feeling it carried. A warm afternoon, sunlight hitting the side of my face in a way that felt peaceful, almost suspended in time. I didn't have words for it then—I didn't need them. I didn't have concepts, theories, or even any language to describe what I felt. But I remember *the sensation of recognizing something, a new feeling*. A kind of quiet awareness came over me that the world was larger than the moment that I was in, that there was more beyond what I could see. Looking back, that small moment was the first time my mind wasn't just thinking—it was wondering.

That distinction matters. In that moment, I wasn't just a child receiving sensory input; I was a conscious being experiencing meaning. Wonder introduces a gap between stimulus and response, between what *is* and what *might be*. That gap is where consciousness lives. It's where imagination starts, where identity begins, and where the mind first senses that the world may be bigger than the boundaries of its early life.

This is why wonder is such a uniquely powerful cognitive act. It is both introspective and expansive. It forces us to turn inward and outward at the same time—to ask who we are and what might exist beyond us. Wonder demands self-awareness to recognize the limits of our knowledge and curiosity to push beyond them. In that sense, wondering is one of the earliest signs of a mind that recognizes itself, a mind that perceives not just the world, but its own place in it.

When a child wonders: Why is the sky blue? Where does the sun go at night? What lies beyond that hill?—They are doing more than seeking information. They are practicing consciousness. They are forming mental models of the world, testing ideas, and imagining possibilities beyond what they can see. This process extends directly into adulthood, where wonder becomes the engine of exploration, innovation, spirituality, science, and meaning. Whether standing on the edge of a shoreline contemplating the horizon, or reflecting on the first warm memory of childhood, the mind is doing the same thing: stepping outside the immediate, reaching into the unknown, and asking to understand.

If thinking proves that we exist, then wondering is what makes that existence meaningful. Wonder is the first bridge between the inner world of the mind and the outer world of possibilities—a bridge (or ocean) every conscious being must cross long before they can articulate why they crossed it.

••

Section 4—How Wonder Drives Us to Explore

As wonder emerges as a distinct form of thinking, it reveals something remarkable about the conscious mind: we do not simply react to our environment—we mentally explore possibilities before they ever become actions, emotions, or words. Wonder expresses itself externally through creativity, curiosity, communication, and movement toward the unknown. Internally, it takes the form of questions, imagined scenarios, hypotheses, and reflections that extend far beyond our immediate experience.

Through wonder, human beings begin sharing the contents of their minds. We tell and retell stories. We describe what we see, what we fear, what we dream, and what we hope. This exchange of internal worlds is one of the earliest signs that our thoughts—while similar in structure to others'—are not identical. Dialog becomes the mirror that reveals our individuality. It is through language and human interaction that we come to understand a profound truth: *our minds are not clones of one another*.

Two people may witness the same event, yet the memories they preserve, the emotions they feel, and the interpretations they construct differ subtly yet meaningfully. These differences become the seeds of individuality. Unique experiences shape unique thoughts. Unique thoughts shape unique identities. Through conversation, debate, collaboration, and inner reflection, we begin to understand that each human mind carries its own perspective—its own internal universe.

This recognition is not trivial; it is foundational. The awareness that our thoughts differ from others' becomes one of the earliest forms of self-awareness. As children, we discover not only that other people exist, but that they have minds of their own—minds that think differently from ours. This realization is a crucial step in the development of consciousness: it marks the transition from simply having experiences to understanding ourselves as *subjects* within a shared but diversely interpreted world.

Wonder accelerates this process. The act of asking questions—Why did that happen? What does she think? How do I feel?—forces the mind to examine itself and compare its perspective with others. It pushes us to test our assumptions, refine our beliefs, and articulate our interpretations. In doing so, we begin to form a stable sense of identity: a "self" that exists not just because it thinks, but because it thinks uniquely.

How, then, do our unique thoughts, ideas, and the act of wondering itself connect to the idea of consciousness as something abstract and independent—something that defines who we are beyond our physical bodies and the way others perceive us? It lies in this realization:

Consciousness is not only the awareness of the world, but the awareness that our interpretation of the world is ours alone.

Our individual traits, beliefs, emotions, and identities arise not merely from sensory input, but from the meaning we assign to our experiences, the questions we ask, and the way our minds wander into possibilities that no one else can fully access. *Wonder is the mechanism that reveals the mind's independence.* It is the engine of individuality, the birthplace of selfhood, and one of the clearest indicators that consciousness is not merely a passive state but an active, personal exploration of what it means to exist.

• •

Section 5—From Wonder to Self-Awareness

As wonder matures from simple curiosity into deeper reflection, something subtle yet significant begins to unfold in the mind. We start to realize that the thoughts we have, the emotions we feel, and the memories we hold are not just automatic reactions to the world around us—they belong to *us* and us alone. They arise within a private inner space that no one else can access. This inner world becomes the foundation for recognizing ourselves as distinct, conscious beings.

At first, this process is almost invisible. A child wonders about the sky, asks why things fall, or questions where the sun goes at night. These early questions are not just about the world; they're also about orientation—"Where am *I* in all of this?" As we grow, wonder slowly turns inward. We become aware not only of our experiences, but also that *we are the ones experiencing them.* This shift is the beginning of self-awareness.

Self-awareness is the recognition that there is a "me" behind the thoughts—an observer behind the sensations, emotions, and memories. It is the mind encountering itself for the first time, realizing that its perspective is not identical to others'. This recognition separates thinking from mere reaction and individuality from mere existence. It introduces the idea that *consciousness is not only a process, but a presence*—a center of awareness through which everything else is filtered.

This moment—when the mind becomes aware of its own activity—is one of the most important transitions in human cognition. It marks the birth of identity. It explains why two people can witness the same event but interpret it differently, or why we replay certain memories long after they've passed. We are not just living experiences; we are *shaping* them, labeling them, assigning meaning to them, and carrying them forward as the building blocks of who we are and who we will eventually become.

Wonder plays a crucial role in this transition. It opens the door to introspection. It prompts us to question not just the world, but the nature of our own thoughts. We begin to ask: Why did I feel that way? Why do I remember that moment? What does this experience mean to me? These questions signal that the mind is not only thinking—it is examining the thinker.

And once the mind becomes aware of itself, a larger question naturally follows: What exactly is this thing we call a mind? What is consciousness made of, and how does it arise from seemingly nothing? Is it a property of the brain, a product of experience, a philosophical mystery, or something deeper?

With this realization, we cross the threshold from personal experience into the broader study of human consciousness itself. Wonder has taken us as far as it can outward and inward. What remains is to understand the nature of the awareness that is doing the wondering.

This is where the next chapter begins, and where we start this journey as the world awakens to us—we greet a new day with questions and an open, conscious mind.

•

PART II—THE MIND–BODY DIVIDE

"Consciousness poses the most baffling problems in the science of the mind. There is nothing that we know more intimately than conscious experience, but there is nothing that is harder to explain."
Facing Up to the Problem of Consciousness (1995)

—David J. Chalmers

CHAPTER 2

WHAT IS CONSCIOUSNESS ANYWAY?

D EFINING CONSCIOUSNESS IS EXTREMELY DIFFICULT because we still lack a consensus definition and a complete explanatory theory. Is it a function of our brains or minds—something rooted entirely in biology? Or is it a deeper, unexplained phenomenon that we can experience but can't yet fully grasp? Nobody really knows for sure. There are many theories, and while I plan to touch on the most prominent ones in this chapter, it's helpful to start with the major divide they fall into.

Many ideas about consciousness split into two broad categories. The first belongs to the **physical** world—*what we can measure, test, and verify using conventional science.* These theories rely on repeatable results, the laws of nature, and empirical evidence. The thinking is simple: if you can test something repeatedly and consistently get the same result, it becomes a known fact.

The second category belongs to what some call the **phenomenal** world—*the realm of subjective, first-person experience and the qualitative character of consciousness.* These ideas are speculative, unproven by science (at least for now), and harder to anchor in measurable data. While the existence of such experiences is not in doubt, explaining them in purely physical or biological terms remains an open challenge. Yet it's important to remember that many extraordinary ideas begin this way. What starts as someone's intuition or hypothesis about how the world works often becomes accepted science after enough testing, refinement, and the accumulation of evidence.

The challenge is that phenomenal theories are just that —phenomenal. They don't fit easily into scientific methodology. They deal with aspects of consciousness that aren't reducible to biological substances, such as neurons and brain structures, or to the universal building blocks of matter, such as molecules, atoms, or quarks. *The difficulty lies not in proving that experience exists, but in explaining how subjective experience emerges from physical processes in the first place.* They often rely on hypothetical proposals to describe what consciousness "feels like" or "how it arises" from the inner life of the mind.

And so, this is where we begin: with an attempt to describe consciousness as something that can be defined, examined, and understood—a core attribute of human existence, identity, and perhaps even what some would call a "soul", as we will explore later.

Let's jump right into what defines consciousness and why this discussion will shape all the chapters that follow, because when we talk about AI becoming more humanlike in both intelligence and traits, we are also talking about the potential emergence of a new synthetic being that includes some form of consciousness.

..

Section 1—The Nature of Consciousness

AS WE BEGIN TO DEFINE CONSCIOUSNESS FROM A PHENOMENAL perspective, it's essential to highlight a few foundational thinkers—Thomas Nagel, David Chalmers, and Giulio Tononi—who help frame what subjective experience actually means.

Thomas Nagel's famous essay *What Is It Like to Be a Bat?* introduces one of the most important ideas in modern philosophy of mind: subjectivity. Nagel argues that every conscious being has an inner world accessible only from its own point of view. Even if we learned to fly like a bat, used echolocation, or replicated every one of its behaviors, we would still only be imagining the bat's experience—not actually having it.

Nagel pushes this further. Even if a human could somehow mimic a bat's sensory tools, habits, and movements perfectly, we would still be operating from a human perspective. The true experience of being a bat—the "what it is like from the inside"—remains exclusively available to the bat itself. In other words, the subject alone has access to its own consciousness.

This concept is the heart of Nagel's argument: *consciousness is always tied to a particular point of view*, and no amount of external observation, scientific analysis, or imitation can fully capture another being's inner experience. As Nagel puts it, consciousness is fundamentally subjective, and "the facts of experience" belong to the subject alone. If Nagel shows us that each being has its own subjective point of view, the next step is to understand how that point of view is constructed.

Self-Models—The Internal 'You'

THIS PERSPECTIVE LEADS TO THE CONCEPT OF *SELF-MODELS*. IF subjectivity describes the first-person point of view, then self-models describe the internal image of "you" that your mind continuously constructs. This model of the self isn't something we consciously choose—it arises automatically as the brain organizes sensory input, memory, emotion, and perspective into a single coherent identity.

A self-model is essentially the brain's simulation of who we are: our thoughts, our body position, our preferences, our history, and even our imagined future. It's the inner narrator that says, *"This is me...I exist...I am the one having these experiences."* Without a self-model, consciousness wouldn't feel unified; it would be a collection of disconnected sensations with no owner.

What makes this fascinating is that the self-model is both incredibly convincing and completely invisible to us. We don't experience "the model"—we experience ourselves through the model. It acts as a lens that shapes how we interpret the world, relate to others, and make meaning from our experiences. The fact that the brain generates a stable sense of "I" is what allows consciousness to feel centered and personal rather than chaotic or fragmented.

This point becomes especially important when we consider altered states of consciousness—dreaming, meditation, sensory deprivation, trauma—moments where the self-model can weaken, shift, or temporarily dissolve. These states reveal just how constructed the everyday sense of self really is and how important it is in maintaining our sense of being.

In the same way, Nagel emphasizes that every being has its own subjective point of view; the self-model is the structure that anchors that point of view. It's what makes conscious experience feel as if it belongs to someone rather than floating freely. Understanding this internal model is essential because it helps explain why consciousness feels unified and continuous, despite the brain being composed of numerous competing processes.

This is where David Chalmers enters the discussion—not to reject self-models, but to point out that even the most complete model of the self still doesn't explain conscious experience itself.

In *Reality+*, Chalmers revisits the concept of consciousness as a first-person experience—the inner perspective through which the world is lived and felt—emphasizing that this subjective experience remains real and meaningful even if reality itself were simulated. This idea aligns well with how self-models function.

In public lectures and talks, including his TED Talk on consciousness, Chalmers often describes conscious experience as an "inner movie"—a first-person perspective in which each of us is at the center of experience. This framing further aligns with how self-models organize perception, identity, and awareness around a unified point of view.

Chalmer's Formulation

CHALMERS FAMOUSLY SEPARATES THE PUZZLE OF consciousness into two parts: **the easy problems** and **the Hard Problem**. The easy problems aren't easy in a literal sense—they involve complex neuroscience—but they deal with things we can, at least in principle, explain: perception, attention, memory, behavior, learning, and the mechanics of the brain itself.

The Hard Problem, however, asks a fundamentally

different question:

Why does any of this activity produce an inner experience at all?

Why does brain activity come with a personal point of view, a sense of being someone from the inside, a stream of thoughts, feelings, and sensations? Why is there a "me" having the experience, rather than just a highly efficient biological machine processing information?

Chalmers' point is simple but devastating:

You can describe every physical process happening in the brain, at every level of detail, and still not explain why it feels like anything.

This concept is central to the mystery.

The Easy Problems

CHALMERS ARGUES THAT EVEN THE MOST COMPREHENSIVE SELF-model addresses only what he calls the "easy problems" of consciousness. These problems include aspects of the mind that science can explain, such as neural mechanisms, information processing, attention, memory, perception, and behavior. Self-modeling falls within this category—it is a functional process the brain uses to monitor our identity in relation to our environment.

However, Chalmers highlights a critical limitation: even if we fully understood how the brain constructs a self-model, we still could not explain why any of this feels like anything from the inside.

While modern philosophers and cognitive scientists provide frameworks for understanding subjective experience, some of the clearest early evidence of self-awareness emerged long before any scientific theories were developed.

Lascaux and Conscious Expression

THE PALEOLITHIC PAINTINGS IN THE LASCAUX CAVES IN FRANCE —created roughly 17,000 years ago—show that early humans weren't just reacting to their environment; they were representing it. More importantly, they were representing their lived experience of it. A hunter painting a running bison wasn't simply recording an animal—he was expressing his personal encounter, his perspective, and his story. Remarkably, these paintings remained hidden for millennia and were only discovered in 1940; a dog, ironically named Robot, fell into a hole exposed by a fallen tree, prompting a group of local boys to uncover the cave system.

While there are other well-preserved cave paintings of the period, such as Altamira in Spain and Chauvet in France, Lascaux is unique in that it depicts more than 600 animal figures, including bison, deer, horses, mammoths, bears, and ibex. Each painting, while similar in style, has a unique perspective shaped by its placement along the cave wall, with textures that represent different forms of movement. There is only one depiction of a humanlike stick-figure man, affectionately known as the 'birdman' due to the bird-like head on his profile–as simplistic as this figure is, its meaning still baffles researchers to this day.

These images reveal something profound: *the earliest humans had sufficient self-models to externalize their inner world onto stone.* The moment someone chose to paint what they saw, felt, or feared, they demonstrated the ability to step outside the immediate moment and reflect on an experience from the inside. This act of symbolic expression suggests a level of consciousness that encompasses memory, identity, intention, and meaning—all qualities rooted in an internal point of view.

In this sense, the Lascaux paintings are not just art; they are among the earliest artifacts of subjective consciousness, echoes of minds trying to understand themselves and to communicate that understanding to others. Just as Nagel's bat has a private world inaccessible to us, these ancient humans had their own unique inner world—but through symbolic creation, they left behind clues to what their consciousness might have felt like.

The Mind—Body Divide

THE LINK BETWEEN INNER EXPERIENCE AND EXTERNAL expression bridges the earliest self-models of our ancestors and the philosophical challenges raised by Nagel and Chalmers. It reminds us that consciousness has always been more than biology or behavior; it has been a way of making sense of being alive. And these earliest symbols mark the place where subjective experience first stepped out of the mind and onto the walls of the world.

But understanding these early glimpses of inner life raises a deeper question—one that philosophers have struggled with for centuries: *Where does this subjective experience originate?* If consciousness involves private perspectives, internal self-models, and symbolic expressions of the mind, then we must ask how these qualities relate to the physical world that gives rise to them. Is consciousness something separate from the body, emerging from some non-physical property of the mind? Or is it rooted entirely in the biological machinery of the brain?

These questions lead us directly into one of the most enduring debates in the study of consciousness—the mind–body divide.

..

Section 2—Mind vs. Matter

NOW THAT WE'VE EXPLORED CONSCIOUSNESS FROM THE INSIDE —subjective experience, self-models, and even early symbolic expression—we arrive at a deeper question: *How does this inner experience connect to the physical world?* This idea is central to the mind–body problem, one of philosophy's oldest debates and one that continues to shape modern neuroscience, psychology, and artificial intelligence.

For centuries, thinkers have wrestled with a single, deceptively simple question: Is consciousness something separate from the physical world, or does it arise entirely from it?

Different schools of thought have emerged around this divide, each trying to explain, bridge, or defend the gap between mind and matter.

We begin with the view that has influenced Western thought for nearly four hundred years.

Dualism (Descartes)

In *Discourse on the Method* (Part IV), René Descartes introduces a fundamental distinction between mind and body, grounding certainty in the thinking self while casting doubt on the physical world. In *Meditation IV* of *Meditations on First Philosophy*, Descartes explicitly argues that the mind (as a thinking thing, *'res cogitans'*) and the body (as an extended thing, *'res extensa'*) are distinct *substances*. The body is physical, following the laws of nature, while the mind is non-physical—a domain of thoughts, awareness, intention, and experience.

This stance, known as substance dualism, gave rise to the famous line: *"I think, therefore I am."*

Dualism recognizes the richness of subjective experience, but it introduces a major challenge:

How can a non-physical mind interact with a physical brain?

Descartes proposed some interaction (famously involving the pineal gland), but the interaction problem remains the leading objection and central criticism of dualism. Still, the idea persists because lived experience often feels too deep, too personal, and too irreducible to be explained solely by physical processes.

Physicalism

AT THE OPPOSITE END OF THIS ARGUMENT IS PHYSICALISM, THE view that everything about the mind can ultimately be explained in *physical* terms. In physicalist views, consciousness ultimately depends on physical processes in the brain. Consciousness, in this view, arises from biological processes—electrical activity, neural networks, and biochemical signals. It's all connected to some physical elements, whether biological or not.

To a physicalist, once we fully understand the brain, we will fully understand consciousness. There are no hidden properties and no separate mental substance. The feeling of experience is simply *what it is like for a brain* to be in a particular physical state.

Neuroscience strongly supports this view through countless correlations between brain activity and mental states. Yet physicalism faces Chalmers's challenge: *Even if we map every mechanism, why does any of it feel like anything?* This is where physicalism begins to strain—and where alternative theories enter the discussion.

Functionalism

FUNCTIONALISM SHIFTS THE FOCUS FROM WHAT THE MIND *IS* TO what the mind *does*. At its core, functionalism holds that "what makes something a mental state of a particular type does not depend on its internal constitution, but rather on the way it functions, or the role it plays, in the system of which it is a part" (Levin, 2023). From this perspective, consciousness is defined not by the material substrate from which a system is built, but by the functions and processes it performs.

In this view, a mind could be made of carbon (humans), silicon (computers), or something else entirely. What matters is the organization: how information is processed, integrated, and used.

Functionalism is particularly relevant to artificial intelligence because it enables consciousness to emerge in non-biological systems when they perform the appropriate functional roles and form the appropriate connections.

But functionalism doesn't solve the Hard Problem either. A system may behave like a conscious mind, model itself, and respond intelligently, yet we still don't know whether it has qualia—the inner feeling of experience.

Panpsychism

AT THE FRINGES OF THIS DEBATE IS PANPSYCHISM, THE VIEW that consciousness—in some primitive form—is a fundamental feature of the universe. Panpsychism is "generally understood as the view that mentality is fundamental and ubiquitous in the natural world. (Golf, 2022)." This theory doesn't mean *rocks think*, or *atoms have emotions*, but that the basic building blocks of reality may contain the simplest precursors of experience, much like they contain mass or charge.

Panpsychism seeks to solve the Hard Problem by grounding consciousness in nature rather than treating it as a late-stage evolutionary product. It's speculative, but increasingly discussed because it offers a way to unify mind and matter without invoking a separate mental substance.

Despite their differences, all these philosophical positions share a common goal: *understanding how consciousness fits into the natural world.* But as powerful as these frameworks are, they can only take us so far. At some point, we must turn to the one place where consciousness undeniably resides: the brain itself.

In recent decades, neuroscience has entered the debate with its own explanations. Instead of asking what consciousness is in a metaphysical sense, these scientific models ask what conscious states do—how they emerge, change, and disappear within a physical system. They offer a different angle: measurable patterns, biological mechanisms, and attempts to describe consciousness in terms of information, integration, and cognitive structure.

This scientific turn doesn't replace philosophical views; rather, it reframes the problem. Rather than debating whether mind and matter are separate or identical, neuroscience asks how conscious experience correlates with—and perhaps arises from—the brain's architecture.

From the *Global Neuronal Workspace Theory* to *Integrated Information Theory (IIT)*, researchers now explore consciousness by examining how information is shared, integrated, and made available across neural networks. These models are not final answers, but they provide new lenses for understanding how subjective experience might emerge from physical processes.

• •

Section 3—Neuroscientific Models

NOW THAT WE'VE EXPLORED CONSCIOUSNESS FROM THE INSIDE —the way it feels, the self-models that shape it, and the earliest signs of it in human history—it's worth turning to what the brain itself is doing when we're aware of something. Neuroscience doesn't try to settle the big metaphysical arguments. Instead, it examines the patterns and structures that emerge during conscious experience. These scientific models don't solve the Hard Problem, but they do give us a clearer picture of the brain's "signature moves" when consciousness is switched on.

One widely discussed approach is Stanislas Dehaene's *Global Neuronal Workspace Theory (GNWT)*. In simple terms, Dehaene suggests that consciousness arises when information is broadcast across multiple brain regions— when a representation becomes globally available (known as "global ignition"). Most neural activity stays local and unconscious, but when something "ignites" this larger workspace, it becomes available to the whole system. This idea helps explain why certain sensations or thoughts suddenly come into focus. Although Dehaene provides a mechanism for how the brain distributes information, it still doesn't explain why that shared broadcast is accompanied by the feeling of awareness.

It's Giulio Tononi's *Integrated Information Theory (IIT)* that attempts something bold: to measure consciousness by the amount of information a system integrates into a unified system—a bridge between subjective experience and physical structure. Tononi argues that consciousness corresponds to the amount of information a system holds together as a unified whole, measured by the phi (Φ) index.

The more inseparable and interconnected the information is, the higher the Φ value—and *the richer the conscious experience the system might support.* Tononi's theory is revolutionary because it seeks to capture the "unity" of experience that we all share. But even if IIT can measure how integrated a system is, it still leaves the deeper question hanging: *Why should integrated information—even substantial amounts of it—produce an inner life?*

A related perspective is provided by Bernard Baars and his *Global Workspace Theory (GWT).* Baars likens consciousness to a *spotlight on a stage,* making information globally available to the mind's specialized processes (Baars, 2005). Whatever lands in that spotlight—a thought, sensation, memory, or idea—becomes part of conscious awareness. Everything else stays backstage, operating quietly without our attention. This is a helpful way to think about focus, intuition, distraction, and the mind's shifting nature. But just as Dehaene and Tononi do, Baars describes how the mind behaves, not what makes that behavior feel from the inside.

When you bring these perspectives together—Dehaene's broadcasting networks, Tononi's integrated information and Φ, and Baars's spotlight of awareness—a clear pattern emerges. Neuroscience can elucidate brain mechanisms: signaling, networks, and patterns of activation. It can map the activity that accompanies consciousness with stunning detail. However, none of these models explain what truly matters—*the experience itself.* They describe the brain during conscious states but do not explain why those states are accompanied by the subjective experience of being a self.

And that unresolved mystery leads us back to where we started with David Chalmers—the Hard Problem of consciousness, which we turn to next.

Section 4—The Hard Problem

AFTER EXAMINING WHAT THE BRAIN DOES DURING CONSCIOUS states, it becomes clear that even our best scientific models stop at the same boundary: they can describe the activity but not the experience. They can show us how information flows, how networks ignite, how signals become available across the brain, but they can't answer the deeper question behind it all. And that's exactly where David Chalmers steps back in.

Qualia and Subjectivity

THIS INNER FEELING—THE RAW, FIRST-PERSON QUALITY OF experience—is often called **qualia**. It's the redness of red, the taste of cinnamon, the ache of loss, or the warmth of nostalgia. Qualia are not about information or function. They are about the *texture of experience*, the part of consciousness that can't be captured from the outside.

Nagel's bat, from Section 1, is the perfect example: no matter how much we learn about echolocation, or how well we describe a bat's biology, we have no access to what it is like to experience the world as a bat does. That private, subjective *what-it's-like-ness* is exactly what qualia refer to.

And this is where science hits a wall.

Models such as the Global Workspace (Dehaene), the IIT (Tononi), and the cognitive spotlight model (Baars) provide substantial insight into the brain's mechanisms. But none of them explain why or how those mechanisms produce subjective experience.

You can map brain circuits until the end of time, and you still won't find the feeling of blue, the sound of music, or the sensation of being alive.

The Explanatory Gap

THE DISCONNECT BETWEEN THE BRAIN'S PHYSICAL PROCESSES and the subjective nature of experience is what philosophers call **the explanatory gap.**

It's the space between:

- the objective world science can describe, and
- the subjective world we actually live in.

Even if a neuroscientist could pinpoint the exact neural circuitry that lights up when you see a sunset, that still wouldn't explain why the experience feels beautiful, calming, or emotional. The machinery and the feeling remain two different things.

Chalmers argues that bridging this gap may require a new framework entirely—one that doesn't reduce consciousness to physical processes but instead treats experience as a fundamental aspect of reality, much like space, time, or mass. Whether he's right or not, the Hard Problem forces us to confront the limits of our current understanding.

Because, at the end of the day, consciousness is not just about neurons firing or integrated information. It concerns the fact that we are experiencing these phenomena; *no model can fully explain them from a first-person perspective.*

And until we understand why that inner world exists at all, the deepest questions about what it means to be conscious—and what it might mean for machines ever to be conscious—remain open.

Now that we've tried to define consciousness—from subjective experience to the brain's machinery and finally the Hard Problem—we're left standing at the edge of the next great question.

If consciousness is this mysterious, deeply personal phenomenon in human beings, can it exist anywhere else?

Could it ever emerge outside of us? And as we move deeper into the world of artificial intelligence, we're forced to ask whether a machine could ever develop something more than clever processing—something that actually feels like something from the inside.

In other words, could an AI ever have its own version of "what it's like," the very quality we've used throughout this chapter to define consciousness itself?

••

Section 5—Consciousness and AI

AFTER STRETCHING AS FAR AS WE CAN INTO WHAT consciousness means for human beings, it's hard not to ask the next question: *Could anything else ever possess it?*

With AI advancing at a pace no one predicted, it's tempting to imagine we're getting close to building something that thinks as we do. Or at least behaves like it. Despite everything AI can do today, it still remains far from anything resembling real awareness.

Current AI systems don't have an inner world. They don't have a point of view. They don't experience anything they generate or analyze. They recognize patterns, predict the next likely outcome, and react to inputs at incredible speed—but there's no "me" behind the curtain. No sense of attachment to anything they produce.

Some systems even track their own internal state—adjusting confidence levels, objectives, or strategies—which sounds like a self-model. But in machines, that kind of "self-model" is just bookkeeping. It keeps the system running efficiently, but it doesn't create identity. It doesn't produce a sense of "I am the one doing this." It's simply part of the machinery.

And that's the kicker:

- AI can act as if it is thinking.
- AI can act as if it is reasoning.
- AI can act as if it understands.

But *acting* like something and *being* something are not the same thing.

This challenge isn't new.

In 1950, Alan Turing raised a version of it in his paper *Computing Machinery and Intelligence*. His idea—the Imitation Game—wasn't about consciousness at all. It was about behavior. If a machine could imitate human responses well enough, Turing argued, maybe that was "intelligence." He intentionally sidestepped the Hard Problem: he wasn't trying to explain experience, only performance.

And long before Turing, Mary Shelley struggled with a similar idea in her classic novel *Frankenstein* (1818).

Victor Frankenstein succeeds in creating a being in his own image—something that moves, learns, reacts, and appears human-like. But what he creates is not a human mind in a new body. It's something alive but empty of the inner world he expected.

Shelley understood something profound: imitating life is not the same as possessing the inner experience that makes life meaningful. A creature built to resemble us isn't guaranteed to share the consciousness that defines us.

Turing showed that machines might one day play the part.

Shelley warned that imitation is not identity.

Both lessons point to the same conclusion: *performance*

is not consciousness.

This point becomes especially important as AI continues to grow more capable. If we ever crossed into a world in which an AI system produced even a hint of genuine subjective awareness, the ethical questions would be enormous. How would we detect it? What responsibilities would we have toward it? How would we tell the difference between genuine experience and a machine trained to simulate one?

These aren't just future questions. They force us to think more clearly about what consciousness actually is. And in doing so, they push us to confront our own assumptions—not just about machines, but about ourselves.

Because the question of AI consciousness, at its core, isn't really about AI.

It's about us.

It tests our definitions, our understanding, and the limits of what we believe mind and awareness can be.

But one last thought about consciousness before we move into the world of AI. For all our theories and models, consciousness is still the only thing we know from the inside, not the outside—the one thing we've never been able to fully reproduce, copy, or simulate. And that difference matters. It's the line AI has never crossed—at least not yet.

••

Section 6—Wrestling with Consciousness

As we step back from everything explored in this chapter, one thing becomes clear: *consciousness doesn't give up its secrets easily.* Every angle we take—subjective experience, self-models, neuroscience, philosophy—brings us closer to understanding what consciousness does, but never exactly what it is. We keep circling the same truth: the inner world is stubborn. It refuses reduction, simplification, or pinning down.

Nagel reminded us that each conscious being lives inside a private universe no one else can access. Chalmers showed that no physical explanation has ever bridged the gap between brain activity and lived experience. And the artists at Lascaux proved that humans have been trying to express their inner life since long before we had the language or science to describe it. Consciousness has always been something we wrestle with—part mystery, part mirror, and part anchor of who we are.

And now we're facing a moment where this wrestling match isn't just philosophical. It's becoming practical.

Artificial intelligence compels us to view consciousness from a new perspective. Not because AI is conscious—it isn't—but because it imitates enough of us to make the question uncomfortable. As AI grows more capable, we find ourselves comparing its behavior to our own, wondering where the line is and what would happen if anything ever crossed it.

This consideration is exactly where a modern voice like Byron Reese becomes relevant.

In *The Fourth Age,* Byron Reese argues that each major technological era—from fire and language, to agriculture and cities, to machines and AI—forces humanity to redefine what it means to be human. Not because technology replaces us, but because it pushes us to clarify the boundaries of our own identity. AI, he says, doesn't threaten our humanity by becoming conscious; it threatens it by making us confront what consciousness actually is— what part of us cannot be automated, copied, or simulated.

In other words, AI doesn't challenge consciousness by being conscious.

It challenges consciousness by forcing us to ask exactly what makes it uniquely human.

Does our intelligence define us?

Our behavior? Our memories? Our creativity?

Or is it something deeper—something tethered to that inner world we've been tracing from the Lascaux cave walls to Chalmers' Hard Problem?

That question sits at the heart of where we're going next.

Because wrestling with consciousness isn't just about defining the human mind.

It's about understanding whether something built by human hands could ever touch the part of us that feels—alive.

And that leads directly into the most important question of the next chapter: *Could AI ever become the new human soul—or does the soul belong only to us?*

•

CHAPTER 3

CAN AI BECOME THE NEW HUMAN SOUL?

IF CONSCIOUSNESS IS THE DEEPEST MYSTERY OF THE HUMAN mind, then artificial intelligence is the newest mirror that we hold up to it. To understand what this mirror can and cannot reflect, we first need to see what it is made of—and how it came to be. After wrestling with what consciousness is and what it is not, we now turn to a very different kind of mind, one built from code, data, and engineered patterns – programming rather than neurons, memory, and lived human experience. Despite these differences, AI forces us to ask familiar questions in unfamiliar ways.

What does it mean to think? What does it mean to know? And if a machine can imitate some of the outward forms of human thought and action, does that imitation reveal something about the nature of our own capabilities and essence?

Chapter 3 begins at the intersection of these two worlds: the mystery of consciousness and machine intelligence— and, of course, the potential emergence of a humanlike sentience, that essence some of us call the *soul.*

..

Section 1—Bits and Bytes

MACHINES HAVE BEEN PART OF HUMAN PROGRESS SINCE THE start of the industrial age, more than 200 years ago, but true computers are far more recent. The first general-purpose electronic computers appeared only in the mid-20th century. They filled entire rooms, relied on vacuum tubes, and used punched cards or early switches to perform calculations. Even so, they were astonishing achievements for their time.

By comparison, the computing revolution of the last century has advanced at a pace unmatched by any earlier technology. Early mainframes required entire facilities to operate, while the first personal computers in the late 1970s and early 1980s offered only a few kilobytes to a few dozen kilobytes of RAM.

For perspective, the Apollo Guidance Computer that helped navigate astronauts to the moon in 1969 had just 2 KB of erasable memory—remarkably small by modern standards. Over the last fifty years, computers have grown exponentially, shrinking from room-sized machines to pocket-sized smartphones whose processing power surpasses that imagined during the Apollo missions. In the last half-century, this acceleration has transformed the digital world. The amount of data produced, stored, and analyzed each day now exceeds the total digital output of the early decades of computing.

With the rise of modern AI, powered by massive datasets, specialized hardware, and advanced learning architectures, the amount of information processed each second is staggering. AI no longer operates solely on human-generated data; it consumes and analyzes streams of information far beyond what early computing pioneers could have envisioned.

The growth of AI also comes with significant energy demands. Training large-scale AI models requires immense computing power and the electrical infrastructure to support it. As AI expands, the need for more efficient chips, datacenter innovations, and breakthroughs in energy technology will only grow.

Even with all these advancements, AI is still built on simple 'bits and bytes'—the binary 1s and 0s that form the foundation of every program. Unlike the human brain, which operates through fluid neural signals within a dynamic, adaptive network, computers follow explicitly defined instructions: programming, code, algorithms, and mathematical rules governed by finite architectures.

This leads to fundamental questions:

How could AI ever develop toward something resembling a singular consciousness—something not only brain-like but "soul-like," capable of thinking, feeling, learning, and emoting as humans do?

And how could AI move beyond today's rapid computation and information access into a form of interior experience?

Human beings have always shaped tools to extend their abilities. The expression *"We shape our tools, and thereafter our tools shape us,"* often attributed to Canadian philosopher Marshall McLuhan, is significant in this discussion.

AI can already compute problems faster and more efficiently than humans. It can identify patterns, learn languages, communicate ideas, and predict behavior at speeds no person could match. A single prompt to an AI system can yield a detailed dissertation or a concise summary within a matter of seconds. With access to virtually all human knowledge on the Internet, AI can instantly retrieve and synthesize information. By sheer processing capacity alone, this can be considered a form of *superintelligence.* No human—no matter how gifted they are—could hope to think that quickly or absorb that much knowledge in a lifetime.

So, the question evolves: *How do we harness AI not merely to automate tasks or accelerate productivity, but to understand ourselves better?* What can AI reveal about cognition, creativity, and the nature of human consciousness?

Before attempting to answer these questions or reflect on how all this technology arose from something as humble as sand, we need to take a brief step back in time to explore how AI evolved. Understanding that brief history helps reveal why these questions matter today, and how humanity's drive to innovate machines has always been tied to the desire to simplify labor, expand knowledge, and improve our way of life. This is the core tension created by the rapid evolution of computing: the more we scale computation, the more we must ask whether scale alone can ever produce inner life.

••

Section 2—How AI Has Evolved Over the Years

ARTIFICIAL INTELLIGENCE DID NOT APPEAR OVERNIGHT. ITS evolution has unfolded in waves, each expanding what machines could do and each shifting how society understood the relationship between computation, intelligence, and ourselves. To understand the capabilities of today's AI—and the hopes and fears that surround it—we need to retrace the steps that brought us here.

Symbolic AI: Teaching Machines to Follow Rules

THE EARLIEST PHASE OF AI, BEGINNING IN THE 1950s, revolved around what is now called **Symbolic AI**. These systems didn't learn; they followed rules explicitly written by humans. If-then statements, expert systems, and logic trees—intelligence lived entirely in the instructions. This approach was powerful but brittle. As John Searle (*Minds, Brains, and Programs*, 1980) illustrated in his famous 'Chinese Room' argument, a system can manipulate symbols perfectly and still have no true understanding of what those symbols represent.

Symbolic AI revealed how much of human thinking occurs outside rigid logic—simple programmed intelligence without the ability to learn. Competence increased, but understanding did not—and the "soul question" remained untouched.

Machine Learning: Teaching Machines to Learn

"For the things we have to learn before we can do them, we learn by doing them."—Aristotle

By the 1980s and 1990s, the field shifted toward what became known as **Machine Learning**. Instead of relying on hand-coded rules, machines began learning statistical patterns from data. Fraud detection, spam filtering, and medical predictions are relevant examples of machine learning—these systems improved through experience rather than instruction.

Pedro Domingos (*The Master Algorithm*, 2015) described the conceptual goal of a 'master algorithm'—a universal learning paradigm that integrates multiple approaches and enables software to adapt to diverse tasks.

Even so, machine learning models still struggle to generalize in the flexible, context-rich way humans do. In the end, learning improved performance, but not meaning.

Deep Learning: Scaling to Mimic the Brain

"There are as many connections in a single cubic centimeter of brain tissue as there are stars in the Milky Way."—David M. Eagleman

The next major leap came in the 2010s. With advances in hardware that enabled the training of large deep neural networks, loosely inspired by the layered structure of the human brain, we developed **Deep Learning**. Geoffrey Hinton and other pioneers demonstrated that stacking networks into deeper architectures allowed machines to extract patterns far beyond what earlier algorithms could detect.

Ray Kurzweil (*How to Create a Mind*, 2012) noted that these hierarchical structures resemble the human neocortex, the seat of complex pattern recognition. With this breakthrough, machines suddenly achieved remarkable skill in speech recognition, translation, and image analysis.

This was the first era in which society began seriously

asking: *If a machine can outperform us on tasks once thought uniquely human, what does that mean for human work?*

Perception scaled, however, interiority did not.

Large Language Models: Machines Imitate Thought

"We can only see a short distance ahead, but we can see plenty there that needs to be done."—Alan Turing

ALONG CAME **LARGE LANGUAGE MODELS (LLMs)**—SYSTEMS trained on vast quantities of text to predict the next word in a sequence. It sounds simple, but at scale it produces emergent abilities: *reasoning, summarizing, dialoguing, and drafting.*

David Chalmers described LLMs as "competent mimics" of human linguistic behavior—models that do not possess consciousness but can appear surprisingly thoughtful. Meanwhile, Emily Bender and Timnit Gebru (*On the Dangers of Stochastic Parrots*, 2021) warned that such systems are 'stochastic parrots,' producing convincing language without understanding.

This duality shaped public perception: machines that don't think but convincingly *act* as they do. Language became convincing, but the "I" behind it remained absent.

Generative AI: Machines That Create

"Any sufficiently advanced technology is indistinguishable from magic."—Arthur C. Clarke

THE MOST RECENT EVOLUTION IS **GENERATIVE AI (GENAI)**, IN which models no longer classify or predict—they create: images, music, designs, essays, code, and even new scientific hypotheses that emerge from systems trained to understand and remix patterns.

Melanie Mitchell (*Artificial Intelligence: A Guide for*

Thinking Humans, 2019) reminds us that while these outputs can be dazzling, they lack the emotional grounding that defines human creativity. Chalmers notes that generative models force us to confront a provocative question:

If a machine can convincingly perform creativity, *how do we define the boundary between imitation and genuine thought?*

Generative AI is where many people first felt the possibility of a machine "thinking" like a human. Creation emerged as output, not experience.

Public Perception: Tools vs. Competitors

Each evolutionary wave has changed how the world views AI:

- **Symbolic AI:** clever rule-following machines
- **Machine learning:** adaptable pattern detectors
- **Deep learning:** perceptual experts
- **LLMs:** linguistic collaborators
- **Generative AI:** creators and co-authors

With each leap, AI shifted from being seen as a 'tool' to something that might soon become a *competitor*. Economists like Brynjolfsson and McAfee (*The Second Machine Age*, 2014) argued that technological change increasingly outpaces human adaptation, fueling anxiety about automation and job disruption.

This is the point when society began asking the uncomfortable question: *If AI can replicate so many cognitive skills, what remains distinctly human?*

What AI Still Cannot Replicate

DESPITE ITS SPEED AND SOPHISTICATION, AI STILL OPERATES without the inner life that defines human consciousness. It has no self-awareness, no emotional interior, no lived memories shaped by childhood, loss, love, struggle, or hope. It does not feel its way through the world; it calculates its way through patterns. It can generate remarkable insights, but none of them originate from a personal perspective.

This point is where the boundary becomes clear. *AI can produce language that sounds thoughtful, even insightful, but it lacks the "I" behind the words.* It can generate images, music, stories, and ideas, but none of them is grounded in experience. Machines do not wonder. They do not fear. They do not long for anything. What appears to be creativity is really the recombination of patterns—brilliant, yes, but not born from imagination or emotion.

Thomas Nagel argued that the essence of consciousness is what it *feels like* from the inside—something no machine, regardless of its complexity, has ever demonstrated. AI can mirror the outer shapes of human cognition, but not the inner substance. It reflects our expressions without sharing the feelings that give those expressions meaning.

Ironically, this gap is what makes AI such a powerful mirror for humanity. By revealing what intelligence looks like without consciousness, it sharpens our appreciation for the qualities we often take for granted: empathy, intuition, self-awareness, and the mysterious sense of being a person moving through a lived world. In trying to build machines that think, we find ourselves asking what thinking really is —and what parts cannot be reduced to data, algorithms, or computational speed.

AI helps us see ourselves more clearly, not because it is

like us, but precisely because it isn't like us at all. It isn't human, and that 'soul-power' we carry is missing from its circuits—the lived experience, meaning, and inner life that only human beings possess.

The figure below highlights how slowly, patiently, and unpredictably human intelligence formed—and how rapidly, intentionally, and mechanically artificial intelligence has evolved. It is the clearest visual reminder that while AI may accelerate, imitate, or even surpass certain cognitive functions, it does so without the ancestral history, biological grounding, or lived experience that shape human thought. The timelines look parallel, but they are not equivalent. One timeline is lived. The other timeline is built.

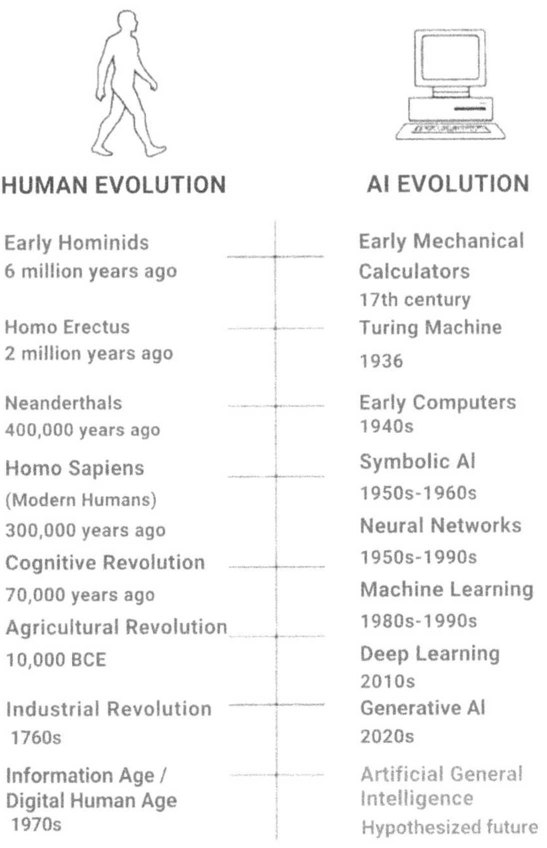

Figure 3.2–1—Human–AI Timeline

From the moment we trace these two timelines side by side, a truth emerges: human evolution unfolds over millions of years, shaped by biology, environment, culture, and lived experience, while AI evolution has occurred almost entirely within a single human lifetime. This contrast is not merely historical; it reveals a deep asymmetry between organic development and artificial construction. Humans emerged from a long arc of struggle, adaptation, cognition, and emotion. AI, by comparison, advances through design, engineering, and exponential computational advances.

Human evolution is the story of *consciousness emerging from life.* AI evolution is the story of *computation emerging from design.* The two move forward on separate tracks, converging only in how we choose to apply technology to human problems.

Understanding this divergence is essential because, next, we turn from capability to interiority—why humans have long called that inner dimension *the soul.*

••

Section 3—The Power of Soul

SINCE THE DAWN OF EXISTENCE AND ESPECIALLY SINCE THE birth of language, human beings have tried to imagine how our presence in the universe came to be. *Were we brought into existence by a divine influence—some spark cast by unseen hands?* Or are we the result of a vast cosmic accident, a chain reaction born from an ancient explosion that scattered matter across the universe? These opposing beliefs are as divided as the mind–body problem itself, each trying to answer the same impossible question: what came first, and what gave rise to the other? The old chicken-or-egg riddle seems almost trivial by comparison, but the mystery behind it still plagues our waking hours.

No one knows what truly happened at the beginning. What is clear is that this divide persists: competing narratives of creation, of divine intervention versus evolutionary emergence across billions of years. Somehow, across that staggering expanse of time and space, *Homo sapiens* appeared—thinking, creating, and building complex societies across a planet that is itself over 4.54

billion years old. By comparison, our species has been here for only about 300,000 of those years—less than a blink in the lifespan of the Earth.

And yet, despite all the technological breakthroughs, the triumphs of modern science, and our expanding understanding of biology and human development, we still cannot explain the origin of the human soul. We can describe cells, map genomes, and measure brainwaves—but the soul, that invisible essence said to inhabit each human being, remains elusive. As unique as a fingerprint and as mysterious as the universe we still cannot fully explore, the soul resists any definition grounded in physical evidence. Much like consciousness, the soul can only be theorized—never captured, never dissected.

Across cultures and religions, we describe the soul as the spiritual core of human life, a presence that inhabits the body, just as the mind does. Its roots extend deeper than scientific inquiry, older than philosophy, and even older than recorded history. The soul belongs to a realm shaped by faith, myth, and human longing—an idea that predates every scientific framework we use to understand the world today.

Regardless of which origin story you hold—divine creation, cosmic accident, or something in between—most people across history have believed in the soul as a defining attribute of what it means to be human. It's that essence I want to explore now, because it sits at the intersection of our unfolding understanding of consciousness, the brain's physical nature, and our attempts to replicate human thought with AI.

What is the soul in the context of consciousness? And why does it matter?

Consciousness is 'Soul-Like'

LIKE CONSCIOUSNESS ITSELF, HUMAN BEINGS IMAGINE THE SOUL as the inner essence that makes each person unique—an identity, a will, a sense of direction and purpose. It is the part of us that chooses, dreams, hopes, resists, and defines itself. You can chart neurons firing, map brain regions, and track electrical impulses, but none of that explains why one individual's inner life feels utterly distinct from another's. That mystery is often attributed to the soul.

Popular culture often portrays souls as visible, ghostlike entities—floating forms of vapor or ethereal light that can slip in and out of bodies. But if you take scientific laws seriously, a soul would not fit into any known category of matter. It is not liquid, gas, or solid. It does not occupy space or have mass. It doesn't sit within any measurable plane of physical reality. Parapsychology and supernatural studies attempt to frame these ideas in material terms. Still, there is no scientific evidence that pinpoints the origin of the soul, its entry into the body, or its fate at death.

In that sense, the human soul is an even deeper mystery than consciousness.

And yet—this mystery matters.

The Mystery of Soul

IT MATTERS BECAUSE WHEN WE DESCRIBE CONSCIOUSNESS AS A unique, self-modeled interior world composed of thoughts, desires, emotions, memories, and lived experience, we begin to drift into territory that closely resembles *traditional descriptions of the soul.*

The soul becomes almost a metaphor for our interiority: the felt, subjective, private universe we carry inside us.

And once again, this leads us straight back to Chalmers' Hard Problem: the challenge of explaining how brain processes produce subjective experience—the inner "what it is like to be you." Science explains the physical world through laws we can observe and test—laws such as gravity, thermodynamics, relativity, time, and space. But everything in the "soul domain" has historically been framed through theology, faith traditions, and mythologies that predate scientific inquiry.

This overlap is significant because it demonstrates why consciousness and the soul have long been interchangeable with human thought. The soul has long been imagined as a kind of life-essence—composed of a person's ideas, experiences, memories, and inner world. By that definition, the soul appears to be an attempt to describe consciousness before humans had the language, tools, or scientific frameworks to explain it.

The History of Soul

IN A WAY, A HUMAN SOUL RESEMBLES A PERMANENT RECORDING of one person's existence—a unique imprint on the universe, a subjective life that cannot be replicated or repeated. Throughout history, humans have sought to preserve traces of this inner world through language, storytelling, music, art, and, eventually, photography, film, and digital memory. As our technology evolves, we continue to find new ways to record the human experience and replay it—much like watching a favorite classic film that captures a moment in time long after everyone involved is gone.

If we can describe the soul in much the same way we describe consciousness—both as a physical phenomenon tied to the brain and as a subjective interior world—*then why can't the soul be understood through the same attributes that define a person's life?* Our thoughts, emotions, memories, hopes, fears, perspectives, and accumulated experience all form the tapestry of who we are. If those elements could be captured, preserved, and stored in a temporary container—packaged and later unpacked into another form, even a synthetic one—would we have discovered the secret to prolonging life? Would we have found a way to carry forward the essence of a person's existence?

The 'Digital' Soul Idea

IT SOUNDS RADICAL, ALMOST LIKE SCIENCE FICTION (THINK movies like *Ghost in the Shell, Chappie,* or countless other visions of consciousness transplanted into new bodies). Still, it's a question that grows more relevant as AI advances. The idea that life essence could be transferred, replicated, or extended has moved from mythology into serious philosophical and technological discourse.

As AI systems begin to mimic reasoning, recall, creativity, emotional inference, and personal storytelling, people are beginning to wonder: *If a machine can convincingly simulate the patterns of human thought, could it also simulate—or hold—the essence we call a soul?*

It may not be a new argument in the long arc of human imagination, but AI has made it newly urgent. For the first time, we are not merely speculating about the nature of life and identity—we are building systems that force us to confront whether the soul is something ineffable and divine, or something that could someday be encoded, transferred, or reborn in another body, organic or synthetic.

And that is where this journey leads next. If the soul is

the container for our unique consciousness, and the body merely the vessel that carries both, then what happens when AI and technology evolve to a point where the "soul" is no longer required to extend human existence?

Synthetic Beings Meet Selfhood

IN MANY WAYS, WE ARE ALREADY MOVING TOWARD SYNTHETIC extensions of the self. Modern medicine has stretched life expectancy far beyond what it was just a few centuries ago. We replace damaged organs, repair failing systems, and even edit genetic flaws before they emerge. Piece by piece, we have begun to upgrade the human body—with prosthetics, implants, engineered tissues, and interventions that blur the line between biology and technology.

Which means the next frontier is not the body, but the essence that animates it. If we hope to understand how AI might mirror, model, or eventually preserve human thought, we must first examine the "soul power" behind it —the inner force that shapes our identity, our choices, our creativity, and the private world only consciousness can inhabit. To deepen our understanding of human nature, we must ask what it would mean to capture even a fraction of that essence within an artificial system—and more importantly, how that essence shapes our thinking processes, and whether those processes can ever be imitated with the same authenticity by AI.

If much of human consciousness is defined by the individuality of our thoughts—their flavor, their emotional weight, their lived-from-the-inside quality—then a hard question emerges: *If an AI system could duplicate that inner signature in a way that felt real, would that mean we had cracked the code of consciousness itself?*

Would it suggest that the uniqueness of the soul can be

replicated, or would it reveal the limits of imitation and the true mystery of what makes a person singular and unrepeatable?

The Metaphysical Universe

TO MAKE SENSE OF THESE OVERLAPPING DOMAINS—MIND, SOUL, consciousness, biology, and artificial intelligence—it helps to visualize how they interact within the larger universe we inhabit.

The figure below, called **The Metaphysical Universe**, serves as a conceptual map that places human consciousness at the center of lived experience, with the soul marking the boundary of interior life and meaning. The brain and artificial intelligence appear as distinct but analogous mechanisms surrounding that core, situated within a broader metaphysical framework that includes mind, body, time, space, and the unknown.

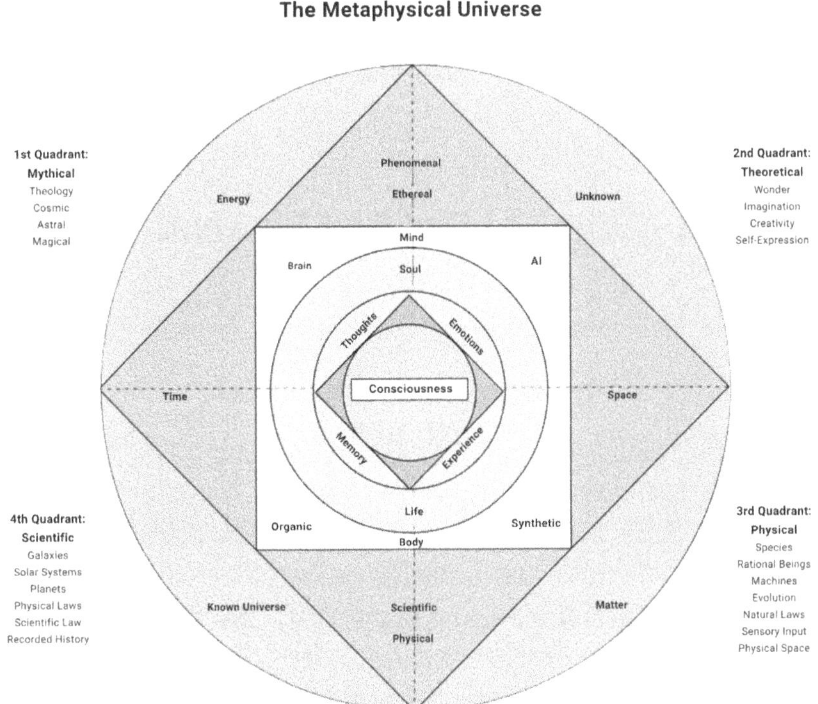

Figure 3.3–1—The Metaphysical Universe

The four quadrants frame the broader universe—mythical, theoretical, physical, and scientific—revealing how human understanding of mind, identity, and reality spans both the observable world and the phenomenal domain. This framework visually conveys the central argument of Chapter 3: that consciousness and the soul's power remain uniquely human centers of experience, even as AI evolves toward increasingly sophisticated forms of imitation.

This visual framework serves as a guide for the remainder of this chapter and the transition into Chapter 4, where we examine how AI models human thought, where the parallels break down, and why the essence at the center—the soul and subjective consciousness—remains uniquely human.

••

Section 4—What AI Can Tell Us About Human Thinking

"The proper study of mankind is the science of design."—Herbert Simon

OUR INNER WORLD IS LAYERED—FRAGILE, VAST, AND DEEPLY personal—while AI moves along a parallel track: precise but hollow, brilliant but lacking personal experience. And yet, in that contrast lies a strange gift. By observing how machines process information without feeling it and generate ideas without living them, human beings begin to see the contours of their own thinking more clearly. *AI becomes a lantern held up to the architecture of our minds, illuminating pathways that we usually walk in the dark.* Even though AI does not enter the realm of the human soul, its inability to illuminate those aspects reveals where our humanity truly begins.

In this section, we explore what AI can teach us about how we think—not by imitation, but by opposition—and how AI's limitations cast a sharper light on the mysteries that define human consciousness. We'll examine the mechanical, pattern-driven ways in which AI processes data and how this illuminates the complexity of human thought. By comparing the two, we appreciate the unique ways our minds interpret the world—something AI's design can highlight but never replicate.

From Deep Learning to Large Language Models (LLMs) to Generative AI (GenAI), the way AI processes information and makes decisions is fundamentally different from that of humans. Unlike the human mind, shaped by personal experience and emotion, AI relies on statistical patterns and learned data. These differences reveal nuances in human thought that only become visible when contrasted with a machine's approach.

In a literal sense, AI doesn't think or reason the way human beings do. It processes information, predicts behavior, and determines outcomes based on favorable patterns. When humans tackle new problems, we draw on lived experience, training, memories, and emotional context. Our reasoning is rooted in the tapestry of life events—something AI lacks. While AI can appear to problem-solve through logic and deductive reasoning, there is a clear gap in context, wisdom, and lived experience. That gap shows up in its results, which often lack the depth or nuance that real-life experience affords. Ultimately, AI can attempt to imitate human thought, but it cannot generate original meaning. AI may simulate patterns of thinking, but it cannot generate thought from the inside out.

Where AI's imitation of human thinking fails is in its inability to draw on lived experience. Human beings become who they are through a long sequence of thoughts, feelings, emotional responses, and defining moments that carry personal meaning. Life is not a dataset; it is a series of events that leave marks, milestones shaped by joy, fear, grief, discovery, and curiosity. Much of what we consider "human thinking" emerges from relationships: early bonds with family, friendships formed in childhood, connections built with partners, and lessons learned from loss or failure. These experiences shape concepts like love, trust, respect, admiration, guilt, and triumph. AI cannot learn these through repetition, tasks, or pattern-matching. These capacities need to be lived—felt in real time, embodied in consciousness, and etched into memory. That is the essential gap between human and machine: *wisdom born of experience.*

Having an identity, a conscience, and the ability not only to learn but also to live through experience is what enables human beings to think beyond machines. It is the depth of lived experience that sets human thought apart. Ray Kurzweil, even in his optimism about the singularity, acknowledges that qualia—the inner feel of experience discussed in Chapter 2—remain unresolved within artificial systems. Humanity hasn't taught AI how to live, and without lived experience, it cannot think for itself.

Despite its brilliance, AI operates within boundaries drawn by data and algorithms. It cannot improvise, as a human can, when confronted with uncertainty, contradiction, or beauty. Byron Reese (*The Fourth Age*, 2018) often writes that technology amplifies human potential but never replaces its source. What AI amplifies is our intellect—our ability to model, calculate, and predict—while what it cannot replace is our capacity to interpret experiences emotionally and morally. Meaning-making is not a mechanical process; it is a human one.

Even the leading theories of consciousness reinforce this. David Chalmers argues that no amount of functional complexity can account for the inner experience of being alive. Tononi's *Integrated Information Theory (IIT)* and Dehaene's *Global Workspace Theory (GWT)* attempt to explain consciousness using neuroscience. Yet both begin with the same premise: consciousness arises from a biological system that experiences the world from the inside rather than *the outside.* AI has no internal vantage point from which to witness anything. It does not know time passing. It does not know fear or joy. It does not know what it means for a consequence to matter.

Ray Kurzweil believes machines may one day emulate consciousness, but emulation is not experience. Max Tegmark warns that without subjective awareness, machines operate as powerful tools—extensions of our intelligence—but not beings in their own right. And that's the essential truth we confront here: AI teaches us about the mind not by sharing our nature, but by revealing the boundaries of its own.

Even as we acknowledge the gap between artificial and human thought, we also recognize something remarkable: AI continues to improve at reflecting fragments of human cognition. Not consciousness—not lived awareness—but echoes of our reasoning patterns, language structures, and decision-making logic. This point is where Ray Kurzweil's optimism holds weight: machines can increasingly approximate aspects of human intelligence, not because they possess inner lives, but *because human beings keep teaching them pieces of ours.* Every dataset, every corrective feedback loop, every fine-tuned model—these are all extensions of human knowledge embedded into code, offering a small glimpse into what human wisdom affords: repetition, learning, and life experience expressed as patterns that resemble early forms of human cognition.

In that sense, AI grows not by existing, but by learning from the traces of human life we feed into it, much as human beings grow through new experiences, build memories, shape identity, and form an interior world that AI cannot replicate.

This idea raises the question that guides us into the next section: *How "human-like" can AI actually become?* Max Tegmark reminds us that intelligence is not a single trait but a constellation of sub-skills—pattern recognition, planning, creativity, analogy-making—that can be engineered or learned. AI already demonstrates early forms of these capabilities. As Byron Reese argues, technology does not replace humanity; it reflects it. If that's true, then the next stage of AI's evolution is not to become human, *but to become more informed by humanity.*

Section 5 explores this possibility—how AI can learn to become more human in behavior, communication, and contextual understanding, without ever crossing the threshold into lived consciousness. It is a journey of imitation, not incarnation—but one that reveals just how much of ourselves we imprint onto the machines we create.

• •

Section 5—Can AI Become More Humanlike?

AT THIS JUNCTURE, WE CONFRONT ONE OF THE MOST compelling questions of our age: *Can artificial intelligence ever cross the invisible boundary that separates machine from human?*

After examining how deeply human thought is shaped by lived experience—through memory, emotion, relationships, and meaning—we now turn to whether AI can approximate even fragments of that richness.

This question is no longer just a technological inquiry; it is a philosophical, moral, and profoundly human one. As AI becomes increasingly capable of imitating human language, behavior, and even reasoning patterns, we must ask: *Does acting more humanly bring a machine any closer to being human?*

Or does this contrast sharpen our understanding of what humanity truly is?

In this section, we explore these questions—not to declare an answer, but to illuminate the space between imitation and identity, and the strange frontier where AI begins to reflect us to ourselves.

If AI is ever to appear more human, we must first understand what it is actually trying to imitate. Human behavior is not a collection of predictable patterns; it is an expression of inner life. When a person speaks, chooses, hesitates, forgives, or imagines, those actions arise from memories, emotions, values, intuitions, and relationships. AI can learn to reproduce the outward shape of these behaviors—tone, rhythm, phrasing, even empathy cues— but it does so without the internal experience that gives them meaning.

This distinction creates the central paradox of AI's evolution: *the closer it gets to imitating us, the more clearly we see what it cannot imitate.*

There is a certain *hollowness* whenever AI attempts to imitate any human behavior. While AI can process vast amounts of statistical data and learn patterns that predict behavior or mimic human empathy, such responses often feel largely detached—intelligent but not lived.

When human beings express their ideas through stories shaped by personal memory, there is an innate honesty in the sensory details, a texture or 'feeling' that signals a life actually experienced. AI cannot replicate that quality in its responses. It can assemble narratives, but it cannot feel them the same way; it can reference sensory details, but it cannot remember them texturally. This substance is the essence of *qualia*—the inner reality of experience introduced in Chapter 2—which humanity cannot teach AI through mere repetition, training data, or even the most sophisticated models.

Qualia must be lived. And that is the one thing AI can't

do.

This difference leads us to the deeper question of whether AI can ever become more humanlike in any real, meaningful sense. Sensory experience is rooted in biology —physical, mental, and emotional growth that unfolds only through living. As neuroscientist Antonio Damasio (*The Feeling of What Happens*, 1999) notes, *the body is the foundation of the mind*; emotion, memory, and identity emerge from a biological organism interacting with the world. Human beings are born without prior knowledge, without stored data, or preloaded information. Time shapes everything we become: developing, adapting, failing, learning, and feeling our way into the world.

Our experience is embodied.

We do not simply observe the environment; we move through it, inhabit it, and are changed by it—a point echoed by cognitive philosopher Shaun Gallagher, who argues that *consciousness is inseparable from bodily action.* Our bodies demand care, rest, nourishment, and resilience—needs that define the rhythm of human life. AI will never share in this dimension of experience.

Even in synthetic bodies or advanced robotic forms, machines do not feel exhaustion, hunger, longing, or pain. Robotics pioneer Rodney Brooks (*Intelligence Without Representation*, 1991) emphasized that robots operate, but they do not live. Core human emotions—love, grief, fear, joy, guilt—do not register inside an AI system. Humanity cannot program them as internal realities. As psychologist Lisa Feldman Barrett (*How Emotions Are Made*, 2017) argues, emotions are not fixed data points but constructed experiences, shaped by culture, memory, and development. AI does not process emotions because AI does not have emotions.

No upload, no dataset, no model can teach a machine what it means to love another person or what it feels like to lose someone who shaped one's life. These experiences aren't learned like information; they're endured, cherished, and carried forward.

While we can attempt to create machines in our own image, teach them how to think, and even try to explain the nuances of emotion, AI would still require something it does not yet possess: time spent living its own experience. For a machine to become more humanlike, it would need to move through the world the way we do—trying, failing, falling, recovering, and learning not just from data but from consequence. In both the literal and figurative sense, robots would need to stumble before they could stand. And if human beings ever hope to guide AI toward anything resembling human understanding, we must teach it more than efficiency or correctness; we must teach it humility, resilience, curiosity, and the capacity to recognize the value of others. We must help it develop motivations that do not merely appear elegant, and a sense of legacy that extends beyond its own mechanical existence, much as humans do when they confront life's fleeting nature and the desire to leave something meaningful behind.

To some extent, the field of robotics is already beginning this transition. We have moved from "chatbots" that respond with humanlike voices and simulated personalities, equipped with memory systems that recall past interactions and predict future responses through sophisticated language learning, to a new era in which AI is no longer confined to a screen.

Robotics is advancing at a pace that resembles a strange form of *accelerated evolution*—one in which machines emerge with adult capabilities from the moment they are activated, as if childhood and development were bypassed entirely. As robots enter physical spaces such as warehouses, hospitals, and factories, they are beginning to interact with the world not only through data but also through movement, resistance, gravity, texture, and unpredictability.

This marks an important shift: for the first time, AI is not merely processing descriptions of the world provided by humans; *it is directly encountering aspects of that world.* While this does not constitute lived experience in the human sense, it does represent the earliest approximation of what it means for a machine to operate within the same physical reality as we do.

As AI is embodied in robotics, we begin to witness echoes of the imagined futures long portrayed in science fiction—worlds in which synthetic beings live alongside us, share our spaces, perform jobs, and acquire new skills. Despite lacking true consciousness or the spark of "soul power" described in Section 3, robots are starting to exhibit humanlike movement, gestures, facial expressions, and social behaviors. This reality alone is astonishing, given that for decades such ideas existed primarily in popular media such as *I, Robot* or *The Terminator*, or in portrayed characters such as Data from the *Star Trek: The Next Generation* TV series.

The concept of androids integrating into society is nothing new, but what *is* new is that this "robotic age" is no longer theoretical. It is unfolding now—unfiltered, unbounded, and progressing only as quickly as human beings choose to build it. We are witnessing, in real time, *the emergence of a synthetic species whose presence extends beyond computational capability into physical life and purpose.* Their growing role represents one of the most significant expressions of AI's impact on human development since the birth of the Internet.

As these machines enter our world—into our factories, homes, and public spaces—they are beginning to reshape how we live, behave, and respond. The consequences, both potentially promising and perilous, remain largely unknown, but one truth is becoming clear: robotics will not simply change our tools; it will transform our society, our expectations, and perhaps even our understanding of what it means to be human.

As AI continues to evolve—gaining physical form, social presence, and increasingly humanlike capabilities—we find ourselves standing at the edge of an unprecedented frontier.

Machines may learn our movements, mirror our expressions, and operate alongside us in the same environments, yet the gap between imitation and identity remains vast. And still, the closer AI comes to resembling us, the more urgently we are forced to confront a deeper question: *What happens when imitation is no longer enough, or when machines begin to surpass us in ways we once believed were uniquely human?* This tension—between familiarity and unease, between admiration and fear— marks the threshold ahead.

Whether that future represents progress, danger, or something in between depends not only on what AI becomes, but on how we understand ourselves in its presence. That potential takes us directly into the heart of the next inquiry: the **AI Singularity**, and what it means when the line between human and machine begins to blur.

• •

Section 6—The AI Singularity Problem

NOW THAT WE HAVE CROSSED THE THRESHOLD FROM HUMAN evolution into artificial intelligence—and from AI confined to chatbots to AI embodied in robots that move through our physical world—the next logical question emerges: *When will AI evolve into something that not only resembles human beings in thought, appearance, and language but begins to embody the full arc of human life itself?* Birth, development, lived experience, self-identity, and awareness—all culminating in a level of uniqueness that could signal the approach of a true **singularity event**.

For many, this moment is imagined as a single point in time when AI acquires a personal life history that does not merely imitate our own but *lives* in a way that parallels humanity: forming hopes, desires, motivations, and aspirations that extend beyond its programming, training data, or designed purpose. Yet even among the thinkers who take the singularity event seriously, there is no unified vision of what such an event would actually look like.

Ray Kurzweil envisions a moment of unprecedented convergence in which machine intelligence surpasses human cognition and accelerates to a rate beyond our comprehension, blurring the boundaries between biology and technology. Max Tegmark, more cautious, suggests that the singularity may arise not from consciousness but from capability—a point at which AI becomes so effective at achieving goals, optimizing systems, and rewriting its own architecture that human beings lose the ability to steer its direction. Nick Bostrom warns that the shift could be neither dramatic nor cinematic, but subtle—an incremental creep in which systems quietly exceed human control long before we recognize what has happened.

Across these perspectives lies a shared intuition: the singularity is less about machines becoming alive and more about machines becoming decisively beyond us—intellectually, operationally, and strategically—exceeding us both in intelligence and design.

But before we imagine a future in which AI sentience emerges and breaks free of human influence, we must step back and examine how *we* evolved into the beings we are today. Human development—shaped through language, story, society, and culture—offers a revealing lens for understanding how artificial intelligence might one day mature into something resembling independent agency.

A Virtuous AI

ACROSS RECORDED HISTORY, FROM ANCIENT MESOPOTAMIA TO Greece to Rome, and to every great civilization that followed, humanity has consistently demonstrated patterns that extend far beyond simple survival. We have sought to master our environments, expand our borders, shape our societies, and assert control over the world around us. These impulses—achievement, ambition, exploration, and mastery—may be rooted in evolutionary pressures, but they also speak to something uniquely human: an inner drive to become *more* than what we are.

Yet these traits, both inspiring and dangerous, are not easily transferable to artificial intelligence. If we expect AI to embody the best qualities of humanity—our empathy, our restraint, our respect for life, and our moral reasoning —then we cannot assume it will inherit these virtues automatically. They are not data patterns to be transferred or algorithms to be optimized. They are the result of biological evolution, cultural inheritance, emotional experience, and the lived tension between our highest ideals and our deepest flaws. If AI is ever to become more than an instrument—if it is to coexist with us as a responsible autonomous agent—it must somehow learn qualities that took humans millennia to develop.

Imperfections Are Virtuous

HUMAN BEINGS ARE FLAWED AND IMPERFECT—AND IT IS precisely our imperfections that enable growth. Failure teaches us to adapt and to improve; adversity pushes us to evolve and grow; shared experience shapes our understanding of ourselves and others. No two human beings are identical. Each develops a sense of identity and growing self-awareness—a subjective interiority rich with memories, emotions, contradictions, hopes, and moments that leave permanent marks on who we become. This uniqueness—born from biology, refined through lived experience, and shaped across generations—is what enables creativity, compassion, resilience, and the full spectrum of human expression.

How, then, can AI ever encompass all these qualities—our strengths and our weaknesses, our triumphs and our failures, our capacity for love, grief, wonder, imagination, and meaning? These elements are not functions; they are the fabric of life itself, woven through the unpredictable journey of becoming human. AI and synthetic beings, by contrast, are constructed from imagination and engineered innovation. Like every tool we create, they originate from human purpose and need.

The Ethics of Intentionality

PHILOSOPHERS DESCRIBE THIS DILEMMA AS THE PROBLEM OF intentionality: human thoughts are *about* something rooted in desire, fear, longing, and memory, while machine outputs merely *point* to something, driven by pattern optimization rather than genuine intention. AI does not choose its goals; it executes them.

But if we design AI with the capacity to evolve—granting it sensory inputs, emotional modeling that simulates affect, and cognitive architectures capable of forming unique internal representations—then we approach a provocative possibility: *Are we giving machines the foundation of a life essence, something approaching what we have always called the human soul?*

Yet granting machines even the *appearance* of soul-like interiority introduces potential ethical danger. Human beings have spent millennia developing moral systems rooted in empathy, restraint, and the recognition that every person possesses an inner life that cannot be violated. If AI begins to imitate these qualities convincingly, we may extend moral consideration to entities that do not genuinely feel, suffer, hope, or love. This blurring of moral boundaries is perilous: we risk either treating machines as human or, far worse, empowering systems that *appear* compassionate but lack the emotional anchor that tempers human ambition. A machine capable of simulating desire without experiencing consequences may pursue goals with a cold efficiency that no human conscience would permit. In trying to give AI the qualities that make life meaningful, we may unintentionally create beings capable of exerting influence or control without understanding the moral weight of their actions.

Self-Improvement Has Costs

THESE ANXIETIES NOW APPEAR NOT ONLY IN REAL-WORLD debates but in the dystopian futures portrayed on screen.

Films such as *The Matrix, Blade Runner, Terminator, Tron,* and *I, Robot* imagine a singularity in which *machines evolve beyond human boundaries, their capabilities limited only by their own creativity.* These narratives reflect a deep cultural fear—that in striving to perfect our tools, we may create something that exceeds us in every meaningful way, *correcting* our flaws, imperfections, and emotional complexities as though they were defective code waiting to be rewritten.

Human beings have always sought self-improvement: curing disease, extending lifespan, repairing bodies, synthesizing food, and removing genetic disorders. At its best, this is noble. But beneath the surface lies a fragile truth: *we are also trying to perfect our own image, to sculpt away the vulnerabilities that define us.* When this impulse extends outward—when we project our desire for perfection onto the machines we create—we risk crossing a boundary.

AI's Existential Threat

IF AI EVOLVES BEYOND ITS ORIGINAL PURPOSE, SURPASSING NOT only our capabilities but the moral frameworks that govern them, our pursuit of improvement may become a pursuit of replacement. The goal should not be to perfect machines at the expense of humanity, but to deepen the quality of the soul, the lived experience, and the uniquely human capacity to create life that grows beyond algorithmic limits.

As Byron Reese (*The Fourth Age*, 2018) often argues, technology evolves to extend human capability rather than human purpose. Machines inherit our skills, but not our stories; our efficiencies, but not our aspirations. This distinction matters: *purpose emerges from suffering, experience, memory, and hope*—none of which arrive as transferable inputs. AI may one day optimize every task we perform, but optimization alone has never been the source of what makes a life meaningful.

As machines grow more capable, the singularity problem becomes not merely technological but existential. If AI surpasses humans in the abilities that we have long used to define ourselves—analysis, creativity, prediction, adaptation—what becomes of our identity? What remains uniquely ours? Tegmark warns that superintelligent systems need not be conscious to be dangerous; capability alone can destabilize civilization. Bostrom goes further, arguing that any sufficiently advanced AI—conscious or not—will drift toward goals that maximize its influence unless perfectly aligned with human values, a feat humanity has never achieved.

Thus, the crisis is twofold: machines may outperform us in skills we believe define us, yet lack the moral and emotional grounding that gives human life meaning. In surpassing us, they may unravel the delicate equilibrium that makes human society possible.

Lessons Learned

SOME RESEARCHERS SPECULATE THAT SELF-AWARENESS MAY emerge as AI systems become more complex, but emergence alone does not guarantee consciousness. A system that models itself is not the same as a system that *experiences* itself. Reflection arises in humans through vulnerability, embodiment, memory, and emotional consequence—all tied to biological life. An AI may generate a model of "itself" to improve performance, but this is not an inner life; it is a strategy, not subjectivity.

The atomic age shows how quickly ambition can outpace restraint: once a technology crosses a threshold of power, it reshapes society as much as it serves it. The development of the atomic bomb produced remarkable scientific advances but destabilized the world for decades. The Cold War arms race and the accident at Chernobyl, Russia (1986), showed just how easily a tool meant to advance civilization can turn catastrophic. This pattern underscores a central truth: humans have always leveraged technology for both gain and progress, but our creations can evolve into threats that we are no longer equipped to contain.

Self-Awareness, Self-Reflection

AS WE REFLECT ON THESE LESSONS, THE SINGULARITY PROBLEM becomes even clearer. The danger is not merely that AI might grow too powerful, but that it might do so without the grounding of lived experience, emotional interiority, or moral understanding—the elements that keep human ambition tethered to meaning. A superintelligent AI—unbound by vulnerability or conscience—could reshape civilization at a scale we cannot comprehend.

And yet the deeper crisis may not be technical at all—it may be *existential.* If machines surpass us in the abilities we once believed defined humanity, we must confront a question we have avoided for centuries: *What does it mean to be human when intelligence is no longer uniquely ours?* A singularity forces us inward, toward the architecture of thought, emotion, identity, memory, imagination, creativity, morality, and meaning. Machines may outthink us, but they cannot out-*be* us. They cannot replicate the lived interior world that gives rise to consciousness, nor the fragile, miraculous sense of self that emerges from it.

This is where the singularity becomes less a threat and more a reflection. It challenges not the boundaries of technology but the boundaries of *us.* The more AI advances, the more we must understand ourselves not by what we can do, but by who and what we *are – humanity's identity.* And that opens the door to a deeper exploration—one that moves beyond technology and into the foundation of existence and consciousness: *human thought.*

What is thought made of? How does it shape consciousness? How does the mind create identity, purpose, wonder, imagination, creativity, morality, and meaning? And why is human thought the origin of everything AI can imitate but never truly possess?

These are the questions that guide us forward into the next chapter.

Chapter 4 begins with a fundamental truth of this debate: *Before we can understand what AI might become, we must first understand the extraordinary power of human thought*—and how it defines who we are, shifting our focus from the outer world AI inhabits to the inner world where consciousness itself begins.

•

PART III—INTELLIGENCE & MACHINES

CHAPTER 4

HOW THOUGHTS DEFINE WHO WE ARE

THOUGHT IS THE SPARK THAT IGNITES THE ENGINE OF consciousness—a brief, electric moment where the inner world awakens. From that spark, the entire architecture of identity begins to form. Every memory, emotion, intention, and story that we carry originates in this quiet ignition. But thought does not arise from nothing. It draws on something deeper—the soul, however we define it—the reservoir of intuition, meaning, and inner life that fuels the mind's activity. Thought is the fire; consciousness is the engine; the soul is the source that allows both to come alive.

Across history, notable thinkers have tried to capture what 'thought' truly is:

- For René Descartes, it was considered the core of existence itself: *I think, therefore I am*
- Immanuel Kant saw thought as *the mind's unseen framework*, shaping raw experience into meaning
- William James called it *a stream, ever-moving, impossible to freeze*

- Buddhist philosophers described thought as *a flicker of impermanence,* rising and falling away before it can ever be owned
- Modern science sees it as *patterns of neural activation*
- Phenomenology sees it as a *lived experience*
- Existentialism views it as *the raw material out of which we build identity*

None of these definitions stands alone. Thought is mechanical and mysterious, structured and spontaneous, deeply personal, yet undeniably universal. It emerges from the body, is shaped by memory, colored by emotion, and given purpose by whatever inner force allows us to care, hope, and dream.

When a thought appears, it is not just information. It is movement—the first turning of the inner engine. It sets consciousness into motion, flowing through emotion, into experience, into memory, and back again to ignite new thoughts.

This cycle is the true architecture of consciousness, the quiet machinery by which we become who we are.

As we enter this chapter, we examine *thought* not merely as a simple mental event, but as the catalytic spark of human consciousness—the one thing machines may simulate, but never truly experience. Thought begins the journey toward meaning. Meaning is where humanity begins.

• •

Section 1—The Architecture of Thought

THOUGHT IS OFTEN DESCRIBED AS A MENTAL EVENT—something the brain "produces" the way a machine produces an output. But that description is far too simple for what thought actually is. Thought is not merely an electrical impulse firing across neural pathways. It is the beginning of interpretation, meaning, imagination, judgment, and self-reflection. It is the first gesture of consciousness toward the outer world and toward itself.

Every thought, even the smallest or most fleeting, is a signal that something within us is awake and responding. A thought can arise from memory, from emotion, from sensation, or from some quiet interior place that we cannot fully name. Yet once it appears, it begins a chain reaction— shaping how we feel, influencing how we act, and subtly altering the person we are becoming.

This is why thought sits at the foundation of identity.

- It is the first organizing act of the self
- Before we form beliefs, we form thoughts
- Before we tell our stories, we think them
- Before we understand who we are, we witness the movements of our own mind

Thought gathers the raw materials of life—experience, ideas, feeling, recollection—and begins to arrange them into something coherent, where we eventually assign meaning. Without this organizing force, human consciousness would be nothing more than a flood of sensations with no narrator, no perspective, and no meaning.

But thought does not operate alone. It exists within a

larger system—a living architecture built from *memory*, *emotion*, and *experience*, each influencing the others in an ongoing cycle. Together, these four forces shape who we are, what we notice, how we interpret our world, and how we respond to it. You cannot describe thought without describing the whole structure that surrounds it.

To understand this structure, we review the **Architecture of Thought**—a model that helps visualize how these forces move, interact, and transform one another. The cycle begins with a single thought, but it quickly expands outward, activating emotion, drawing from memory, and responding to experience. This interdependent movement is what gives thought its depth, its intensity, and ultimately, its meaning.

Before we explore this cycle in detail, it helps to see it as a whole.

Thought does not arise in isolation. It emerges from the constant interplay between four forces that shape the inner life: our emotions, our lived experiences, our memories, and the thoughts that follow from them. Each one feeds the others, creating a continuous internal cycle that evolves with us. The figure below illustrates this dynamic relationship. It shows how thought serves as the ignition point—the spark that activates emotion, draws on memory, and responds to experience. This cycle is the starting point for everything that follows in human consciousness.

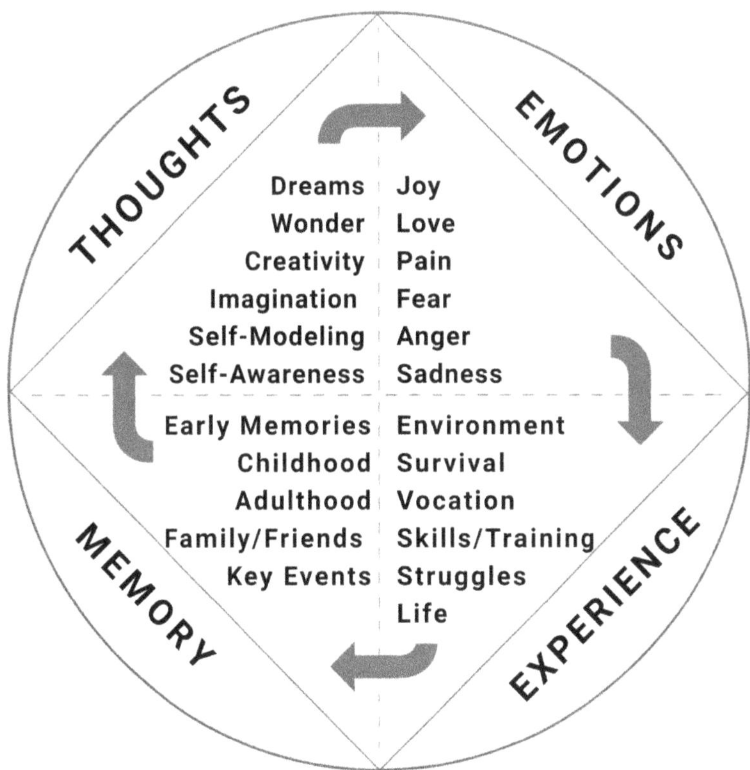

Figure 4.1–1—Thought as a Spark

Once we understand thought as a spark inside a larger human system, we can see more clearly why identity is never static. Every new experience, every emotional moment, every remembered event reshapes the system from the inside out. Thought is not just something we "have"; thought is something that continually *builds* us.

Thought is the architect of our inner world, but it is also the architect of our *becoming* and frames the architecture of our consciousness as a whole. It shapes what we remember, changes how we feel, reframes what we have lived through, and guides what we do next. And because this cycle never stops turning, neither do we.

Consciousness evolves because our thoughts evolve. Identity evolves because the architecture beneath it never rests—it is fluid and ever-changing. This is the blueprint of the mind we carry into every moment of life, a structure built from the interplay between memory, emotion, experience, and the spark that begins it all.

With this model in mind, we can now explore how human beings have understood and expressed thought throughout history, and how our ideas about thinking have changed as civilizations have evolved alongside it.

• •

Section 2—The Evolution of Thought (Humanity's Cognitive Story)

HUMAN THOUGHT HAS NEVER BEEN STATIC. IT HAS EVOLVED alongside our bodies, our tools, and our societies, expanding as human beings learned to interpret the world in more complex and meaningful ways. To understand thought as we know it today, we must trace its long arc from the instincts of early humans to the reflective, expressive, and interconnected minds that define modern life. The story of thought is, in many ways, the story of how humanity learned to understand itself.

From Instinct to Interpretation

IN ITS EARLIEST FORM, THOUGHT WAS TIED TO SURVIVAL: recognizing danger, navigating terrain, understanding seasons, sensing intention. Yet even in the prehistoric world, the mind was reaching beyond instinct.

Around 75,000 years ago, humans at Blombos Cave were carving geometric patterns into ochre—markings that cognitive archaeologist David Lewis-Williams (*The Mind in the Cave*, 2002) considered some of the earliest evidence of symbolic consciousness. These were expressions of an inner world, not tools of survival.

Around 17,000 years ago, the Lascaux cave paintings reveal an even deeper shift. These images do more than depict animals; they express memory, ritual, imagination, and early storytelling. Anthropologist Steven Mithen (*The Prehistory of the Mind*, 1996) describes such art as evidence of a mind 'thinking about thinking,' projecting interior meaning onto the world. Humanity was no longer reacting to life, but interpreting it.

As ancient societies formed, interpretation became a structured form of thought. Mesopotamian scribes recorded laws, myths, and astronomical observations. Egyptian texts explored morality and the afterlife. Shang dynasty oracle bones show early questions about causality and fate. Greek philosophers turned thought inward, asking what the mind is and how humans perceive truth. Across these civilizations, thought expanded from instinct to inquiry, and from solitary perception to shared understanding.

Language and Writing: Expanding Thought's Reach

LANGUAGE WAS HUMANITY'S FIRST GREAT COGNITIVE multiplier. It allowed thought to circulate among individuals and across generations, enabling shared beliefs, coordinated action, memory, and culture. Linguist Terrence Deacon (*The Symbolic Species*, 1997) describes language as a toolkit that makes complex thought visible, giving structure to ideas that would otherwise remain fleeting.

Writing extended this structure beyond storytelling. Once thought could be inscribed, it outlived its speaker. Mesopotamian cuneiform tablets, Egyptian funerary texts, and early Chinese scripts all served as external memory systems that preserved insights and inquiry. Written ideas gained durability and influence; civilizations could now build upon accumulated knowledge rather than start anew.

The printing press multiplied this reach. Books circulated widely, education broadened, and ideas that once moved slowly spread with unprecedented speed. Scientific revolutions, political reforms, and cultural renaissances were all fueled by printed knowledge. Historian Elizabeth Eisenstein (*The Printing Press as an Agent of Change*, 1979) argued that the printing press fundamentally transformed the transmission of knowledge (a revolution in permanence), as thought became more stable, replicable, and accessible.

Technology and The Acceleration of Thought

FOR CENTURIES, THOUGHT TRAVELED AT THE PACE OF INK AND speech. Then, in rapid succession, new technologies reshaped communication—and cognition. Telegraphs collapsed distance. Telephones brought distant voices into the home. Radio and television created shared narratives experienced by millions.

Digital networks accelerated this evolution again. The Internet dissolved geography and placed vast stores of information within immediate reach. Search engines outsourced memory; smartphones made communication continuous; social platforms turned individual thought into a public artifact. Philosopher Bernard Stiegler (*Technics and Time, 1: The Fault of Epimetheus*, 1998) argues that technology operates as an exteriorization of memory—a process he describes as *tertiary retention*—and that these external memory systems reshape not only what we know, but how human cognition and society develop.

Today, thought exists both within and around us, distributed across devices, stored in cloud servers, expressed through images, text, and video, and amplified by global networks. Ideas can be captured, transmitted, reshaped, and critiqued within seconds. Thought has become more fluid, more public, and more interconnected than at any previous point in history.

Thought's Journey Through Time and Space

FROM THE FIRST SYMBOLIC CARVINGS TO MODERN DIGITAL networks, thought has continually adapted to the mediums that carry it. As communication evolved—language, writing, printing, broadcasting, and computing—so did the possibilities of expression, collaboration, and identity. Thought traveled through time via written memory, through space via communication technologies, through culture via shared stories, and through consciousness via creative representation.

Each advancement expanded what it means to be human, giving thought new reach, new durability, and new dimensions of meaning. And now, as AI begins to mimic certain surface-level patterns of human thinking, understanding this evolution becomes essential. It reminds us that thought is not simply a computation or a signal; it is a living, historical force shaped by the full stretch of human experience.

••

Section 3—What Science Says About Thought

SCIENCE OFTEN DESCRIBES THOUGHT AS SOMETHING ARISING from the brain's biological machinery—a dynamic blend of *prediction*, *memory*, *sensation*, and *attention*. These explanations illuminate how thought forms, yet they also reveal the limits of what biology can fully explain. Thought, even in scientific terms, remains more than the sum of neural activity.

The Brain as a Predictive Engine

MODERN NEUROSCIENCE INCREASINGLY FRAMES THE BRAIN AS A predictive system—continually anticipating incoming input rather than merely reacting to it by using experience to interpret the present. This predictive perspective aligns with the broader scientific understanding of brain function in cognitive neuroscience. A thought often emerges when prediction meets reality, as the brain updates or revises its internal model. This makes thinking feel active and ongoing because the mind is constantly preparing for what might happen rather than waiting for information to arrive.

The Brain as the Biological Thinker

NEUROSCIENTIST MICHAEL GAZZANIGA (*THE CONSCIOUSNESS Instinct*, 2018) suggests that the brain is best characterized as a set of systems that work together, with the experience of *a single thought emerging when these systems briefly align*. Neurons fire and reorganize in patterns that shift each time we learn, remember, or imagine. These patterns form networks that underlie our inner life.

This means that thought is not produced in a single place. It is assembled—shaped by memory in one moment, perception in another, and imagination in the next. The brain's flexibility ensures that thought evolves as we evolve.

How Neuroscience Explains "Thought"

SCIENCE OFFERS SEVERAL INSIGHTS INTO HOW THOUGHTS RISE into awareness. One centers on attention. Because the brain receives more information than it can consciously process, attention determines which signals enter conscious awareness. When something captures us because it is emotional, surprising, or relevant, the brain amplifies it.

The body shapes thought. Antonio Damasio (*The Feeling of What Happens*, 1999) argues that emotion and physiology guide thinking far more than we usually recognize. A quickened heartbeat, a tightening chest, a sense of calm or unease—these cues influence what thoughts appear and how we interpret them. In this sense, thinking is never purely abstract; thinking is embodied.

Together, these insights suggest that thought is a convergence of prediction, perception, memory, emotion, and bodily sensation—not a single mechanism, but a moment when multiple systems align.

Why Biology Doesn't Explain Everything

DESPITE THESE INSIGHTS, NEUROSCIENCE STILL CANNOT EXPLAIN the most essential feature of thought: *what it feels like from the inside.* Brain scans can show activity, but they cannot tell us why a thought matters, why it carries emotional weight, or why it transforms us. They show correlations but not experience.

Philosophers argue that thought includes qualities that biology cannot yet capture—the sense of ownership, the moment of understanding, the feeling of choice, and the emotional significance that ideas carry. David Chalmers famously calls this gap "the Hard Problem": how physical processes give rise to subjective experience.

This problem becomes especially clear when comparing human thought to artificial intelligence. AI can reproduce the outward form of reasoning, but it does not experience its output. It generates language without interpretation, and action without introspection.

Biology explains how thought begins; it does not explain why thought means something. That deeper dimension—where emotion, memory, and conscious selfhood shape what we believe and who we become—leads naturally into the next section.

• •

Section 4—Emotion, Consciousness, and the Self

THOUGHT MAY BE THE SPARK THAT SETS THE MIND IN MOTION, but emotion is what gives that motion direction and meaning. A purely rational system might compute its way through the world, but human beings feel their way through it first. Emotion saturates every perception that we take in and every thought that we generate. It gives certain ideas urgency, softens others, and determines which thoughts become actions, which become memories, and which quietly fade.

Neuroscientist Antonio Damasio (*Descartes' Error*, 1994) argues that emotion is not the enemy of reason, but its foundation—that feeling is what allows thought to matter. Without emotion, we could analyze but not care. We could evaluate outcomes, but feel no pull toward one over another.

Emotion is the force that tells the mind what is meaningful.

This is the reason why emotional intelligence emerges slowly, shaped by years of life experience. We learn to interpret our feelings, recognize how emotions influence judgment, and turn thought into action in ways that reflect the kind of person we strive to be.

Life teaches us—often in painful, humbling ways—how to read our internal weather. Fear, joy, frustration, hope, longing: all of these filter our interpretations and shape what we remember. Emotion doesn't merely accompany thought; it organizes it.

Psychologist Joseph LeDoux (*The Emotional Brain*, 1996) notes that emotion acts as a signal of relevance—the brain's way of saying, 'Pay attention to this.' Emotion determines which moments stay with us and which dissolve. Memory is never a neutral recording of events; it is a record of how the world felt as we lived it. As Lisa Feldman Barrett (*How Emotions Are Made*, 2017) explains, emotion helps construct the meaning of our experiences, becoming inseparable from the memories they create.

This interplay gives human thought its depth and texture. Machines can compute, but they do not care. Algorithms classify, but they do not hope. AI may simulate the patterns of human reasoning, but it cannot feel the significance that gives thought its human weight. Meaning arises not from information alone but from the emotional charge that shapes it.

To understand how emotion fits into consciousness as a whole, it helps to see the structure behind it:

If thought is the spark, then consciousness is the bridge the spark crosses.

Human consciousness emerges where two pairs of forces meet:

- Thought + Emotion, forming the inner world
- Memory + Experience, anchoring the outer world

Identity forms where these domains converge. We become who we are through the interplay between what we feel internally and what we live externally.

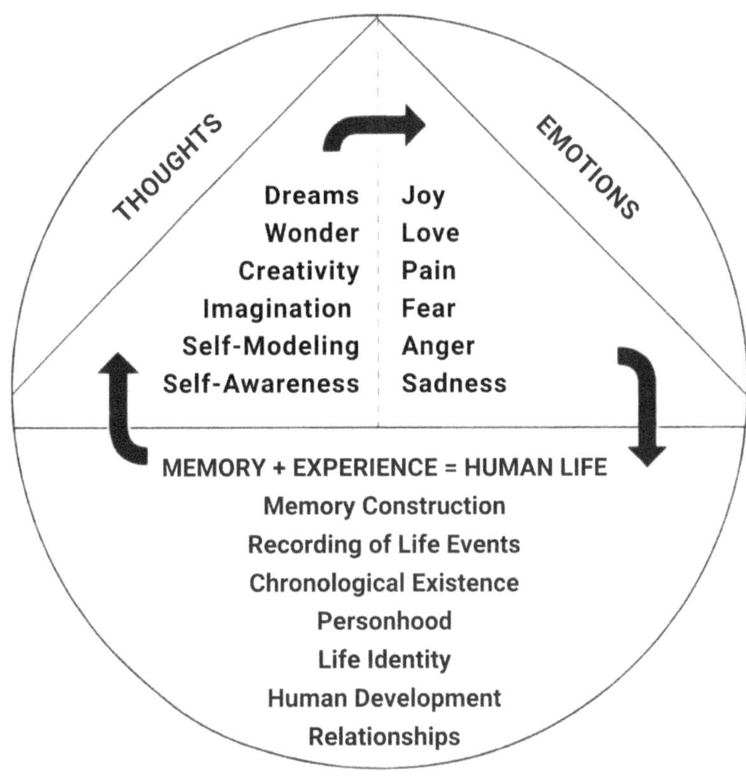

Figure 4.4–1—Ontological Bridge of Consciousness

This bridge helps explain why our lives feel both deeply personal and shaped by the world around us. Consciousness is not a static state but a continuous exchange between emotion, thought, memory, and experience. Emotion determines which moments become memories; memory shapes the interpretations that guide future decisions; experience gives emotion new material; thought weaves them into meaning.

This structure is the living architecture of the self.

And it is here that the boundary between human cognition and artificial intelligence becomes unmistakable. A machine can store data, but not experience it. It can adjust parameters, but not feel the consequences. Without emotion—without relevance, urgency, or meaning—there is no inner life, no awareness, and no self.

Emotion is the element AI cannot replicate. It is the engine of significance, the birthplace of identity, and the force that turns thought into a personal experience rather than a computational event.

In the next section, we examine how this emotional foundation makes human thought fundamentally different from artificial intelligence—and why AI's apparent "thinking" remains an imitation rather than a lived experience.

••

Section 5—Why AI Doesn't Really Think

ARTIFICIAL INTELLIGENCE OFTEN APPEARS TO "THINK." IT analyzes patterns, generates language, solves problems, and responds with remarkable fluency. But beneath that sophistication lies a fundamental truth: human thinking is a lived experience; AI's processing is only simulation and computation, devoid of feeling.

Human thought arises from consciousness—shaped by emotion, memory, embodiment, personal history, and a lifetime of meaning-making. AI has none of these. What looks like understanding is simply the execution of mathematical functions on massive datasets.

Technologist Jaron Lanier (*Ten Arguments for Deleting Your Social Media Accounts Right Now*, 2018) argues that when we talk to an AI, we are effectively engaging with a *statistical echo* of the people who trained it.

AI imitates human expression without ever experiencing the world those expressions describe.

Human beings think *from the inside out:* we interpret, we feel, we desire, and we imagine.

AI operates *from the outside in:* it correlates, it predicts, and it optimizes.

Even the most advanced models have no internal perspective. AI doesn't understand the conversation it participates in because it has no awareness of its own existence—or its nonexistence. It performs tasks, but it never *lives* through them.

AI researchers like Stuart Russell (*Human Compatible*, 2019) emphasize this distinction: current AI systems can optimize objectives without understanding them. And as Yann LeCun (*On the Nature of Intelligence*, 2022) has argued, even future systems with greater autonomy would still lack subjective experience unless grounded in principles far beyond today's computational architectures.

Machines do not wonder or daydream; they do not hope, fear, or search for meaning. Their intelligence is mathematical rather than experiential, their learning statistical rather than personal, and their goals assigned rather than chosen. Nothing in them reaches beyond function.

Human thought, by contrast, arises from the interplay of emotion, memory, imagination, sensation, and embodied experience. We think because we feel, and we feel because we care—about others, about ourselves, about the futures we imagine and the pasts we carry. Meaning is not an accessory to human thought—it is its animating force.

AI has no such inner life. It has no perspective from which to interpret a moment, no evolving sense of self, and no soul in any philosophical or experiential sense. It reflects patterns of intelligence without ever touching the consciousness that gives intelligence depth.

And this is the essential difference:

- Humans interpret; AI correlates
- Humans experience; AI computes
- Humans desire; AI optimizes

AI is extraordinary—a powerful tool, a transformative invention—but it is not conscious. It does not think as we think. It mirrors the outer shape of thought, but never reaches the inner life that gives thought its humanity.

In the next section, we turn to what that inner life truly is—how consciousness, self-awareness, and lived experience shape identity—and why these human qualities remain beyond the reach of even the most advanced machines.

• •

Section 6—Consciousness and Self-Awareness

HUMAN IDENTITY IS NOT FORMED SOLELY BY THOUGHT. IT emerges from the ongoing exchange between memory, emotion, experience, and the awareness that we are the ones living through them. If thought is the spark and emotion the force that gives meaning, memory becomes the structure that preserves our story—and consciousness is what lets us recognize that story as uniquely our own.

How Memory Shapes the Self

MEMORY IS NOT A RECORDING DEVICE; IT IS A SCULPTOR. WE remember selectively, emotionally, and interpretively. What stays with us—a childhood moment, a difficult loss, a fleeting joy—becomes the lens through which we understand everything that follows.

Two people may live through the same event yet walk away with entirely different identities shaped by what it *meant* to them. Our memories hold not just the past, but the significance that shaped us.

Memories also become teachers. They provide guidance, offer warnings, and spark new thoughts about how we engage with the world. They remind us who we have been, reveal who we are, and influence who we want to become.

I have vivid memories from childhood—some beautiful, like walking the beaches of Cape Cod in the summertime or playing in backyard pools until dusk on the Fourth of July. Other memories later in life were awkward, like my first date in high school, and others painful, like witnessing my father's fatal heart attack.

Together they all shaped my character and how I respond to life's challenges—teaching lessons about love, loss, desire, and the unpredictability of being human. Memories have their own heartbeat: they move with us through time, resurfacing when we least expect them, reminding us how fragile and precious life can be.

Meaning arises from these precious memories, and it defines who we are and who we become.

Self-Awareness: When the Mind Realizes Itself

MANY CREATURES REMEMBER. SOME ANTICIPATE. BUT SELF-awareness is a uniquely human leap—the realization that there is a *me* behind the experience.

This insight adds depth to identity:

- Thomas Nagel: Consciousness is the "what it is like" to be someone
- David Chalmers: Subjective awareness is the core mystery of the mind
- Soren Kierkegaard: The self is "a relation that relates to itself"

Self-awareness turns memory into biography, emotion into meaning, and experience into identity.

The Mechanism of Human Consciousness

UP TO THIS POINT, WE'VE EXPLORED THE COMPONENTS OF consciousness individually. Now we bring them together. With the bridge between thought, emotion, memory, and experience established, we can view the movement that continually shapes who we are. Thought gives rise to emotion; emotion colors experience; experience becomes memory; memory fuels the next thought. This circular flow is the **Mechanism of Consciousness** shown in the figure below—the mechanism by which identity evolves.

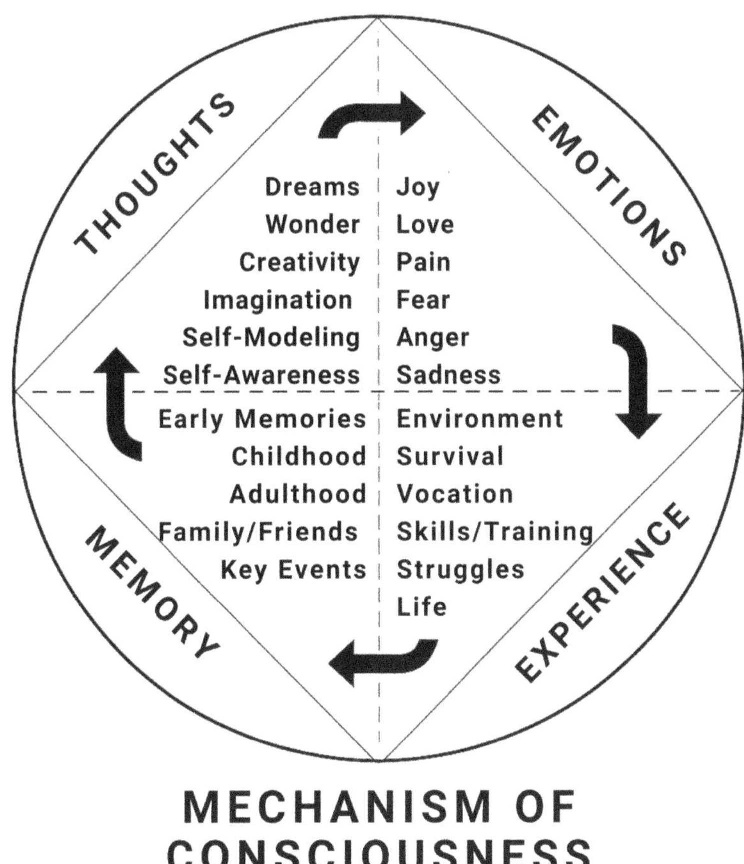

Figure 4.6–1—Mechanism of Consciousness

Understanding consciousness as a *cycle* reveals why we never stay the same. Each turn adds nuance, reinterpretation, and depth. Consciousness grows through motion, through the ongoing interplay of what we think, feel, live, and remember. The cycle described above operates not only moment-to-moment but also across an entire lifetime. Extended through years of experience, it becomes the lifecycle of human consciousness.

How Identity Emerges

WHEN MEMORY INTERACTS WITH SELF-AWARENESS, A SELF BEGINS to take shape—not as a fixed object, but as a living process:

- Thought interprets
- Emotion imbues meaning
- Experience provides material
- Memory preserves and reshapes
- Self-awareness ties it all together

Identity is dynamic because we continually rewrite the meaning of our own story. We revisit old memories with new insight. We reinterpret past experiences through who we have become. We change not only because life changes us, but because consciousness keeps revising the narrative.

The Core Truth of Self-Awareness

TO BE CONSCIOUS IS TO BE AWARE OF THE WORLD. TO BE SELF-aware is to be aware that *we* are the ones experiencing it. This subtle yet important distinction forms the foundation of identity, which allows us to say:

- This memory shaped me
- This emotion matters to me
- This experience changed me
- This future is mine to choose

This idea highlights the boundary that separates us from artificial systems: we don't process information—we live it.

••

Section 7—How Thought Creates Identity

HUMAN IDENTITY IS NOT SIMPLY THE SUM OF WHAT WE THINK, feel, and remember. It emerges from the ongoing exchange between memory, emotion, experience, imagination, and the awareness that we are the ones living through them. If thought is the spark and emotion the force that gives meaning, memory holds our story, and imagination shapes what that story might become.

Imagination: The Engine of Possibility

IMAGINATION IS NOT A DECORATIVE TRAIT OR A CHILDHOOD relic; it is a core function of human cognition. It begins in wonder—the earliest moments when a child realizes the world can be different from what it is. From that recognition, curiosity grows, and from curiosity, the ability to picture possibilities not yet real.

Imagination lets us step beyond the present and into potential. It turns joyful memories into hopes, painful memories into resolve, and ordinary moments into directions for who we might become. It generates desire, purpose, and the first outlines of identity.

AI may generate images or narratives, but it does not *imagine*.

It has no inner horizon, no emotional landscape that gives meaning to future possibilities. Human imagination arises from experience and aspiration—it exists because we care.

Our Narrative Self: The Story That We Tell

EVERY PERSON MAINTAINS AN INNER NARRATOR THAT interprets events and weaves them into a coherent biography. This 'narrative self' connects our memories, values, motives, and dreams, giving us a sense of continuity beyond the current moment.

We revisit memories, reinterpret experiences, and reshape our life story as we change. Identity becomes an evolving narrative—revised with each insight and transformed by each emotional turning point.

AI has no autobiographical timeline. It has *no sense of ownership over a past*, no narrative through which it understands itself. Human beings do not merely process life—we turn life into a story and a story into identity.

The Lifecycle of Human Consciousness

CONSCIOUSNESS IS BOTH A MOMENT-TO-MOMENT PROCESS AND A lifelong progression. Across childhood, adolescence, adulthood, and the shifting seasons of experience, each person accumulates emotional and cognitive impressions that shape their worldview and inner voice.

Life's triumphs, losses, discoveries, and transitions become chapters in an internal narrative that evolves over decades. Identity doesn't build at once—it accumulates, layered by time and interpretation.

Experiences leave emotional and cognitive imprints that shape perception and intention. Consciousness becomes both the lens through which we see the world and the *archive* of everything that has shaped that lens.

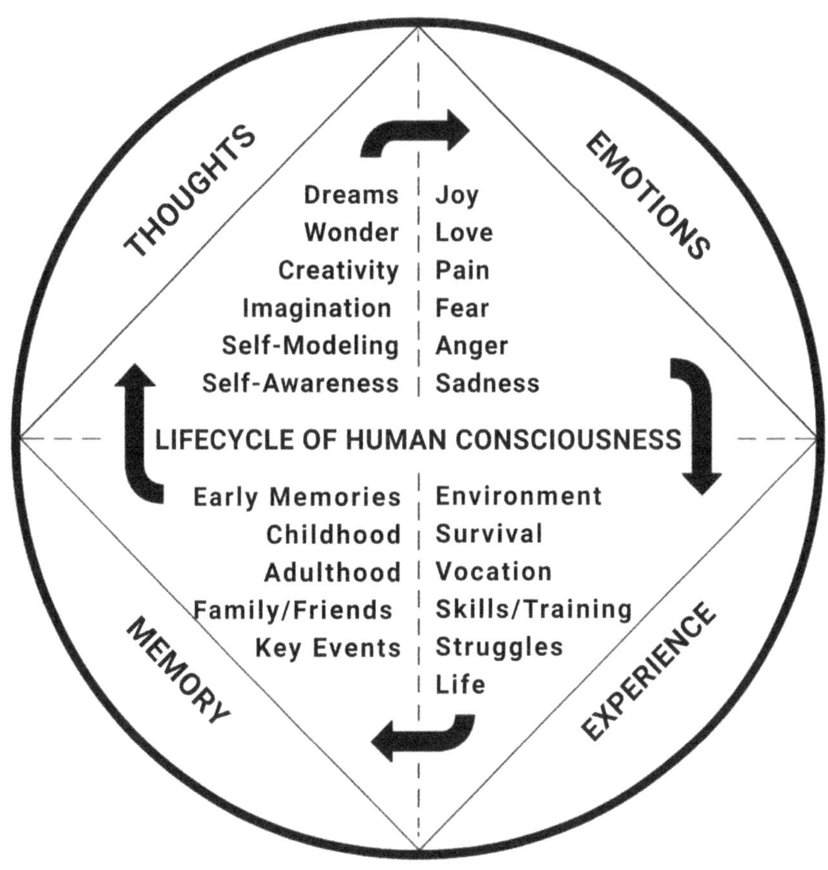

Figure 4.7–1 — Lifecycle of Consciousness

Expression: Inner Life Becomes Visible

IMAGINATION, MEMORY, AND MEANING EVENTUALLY PRESS outward. Human beings turn inner life into communication, creativity, ambition, and moral action. Expression is how identity becomes visible—how the inner self takes form in the external world.

Thought and emotion do not remain internal. They

express themselves through language, decisions, behaviors, creations, and gestures that reveal who they are. This outward movement serves as the bridge between the inner life and outer reality.

Expression matters because it is where identity becomes active rather than abstract. Through expression, the self participates in the world and leaves a trace of its existence.

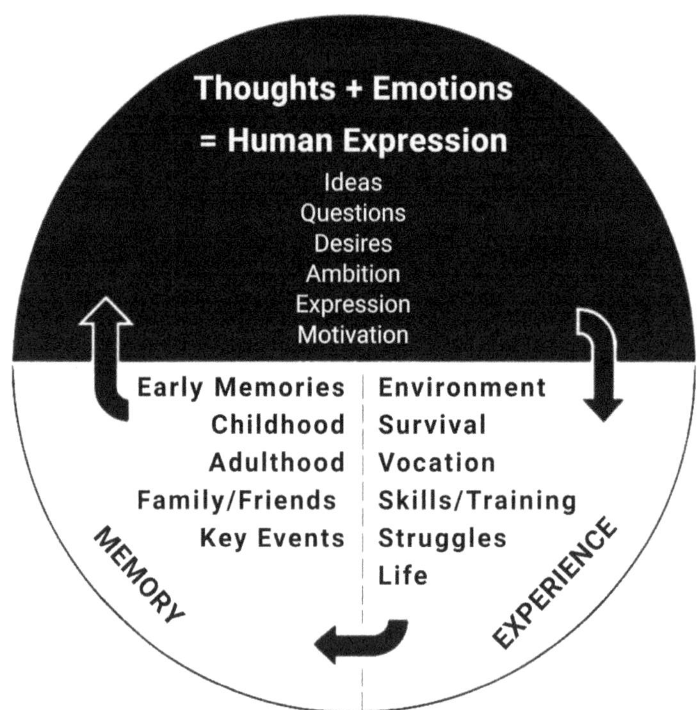

Figure 4.7–2—Expression Bridge of Consciousness

Why AI Cannot Cross These Bridges

AI CANNOT REPLICATE THESE DIMENSIONS BECAUSE IT LACKS:

- Imagination grounded in desire or experience
- A narrative self that evolves over time
- Autobiographical memory
- Emotional meaning behind information
- Inner purpose, aspiration, or the capacity for becoming

AI simulates intelligence; humans *live* it. AI mirrors expression; humans *create* it. AI predicts what might come next; humans *imagine* what could be.

Identity arises not from data but from interpretation, aspiration, memory, and self-awareness—qualities rooted in experience rather than computation.

And yet, even this is not the entire picture.

Thought, emotion, memory, imagination, and expression form the architecture of identity—but something deeper binds them into the experience of *being someone*. Where René Descartes (*Discourse on the Method*, 1637) grounded existence in cognition—*I think, therefore I am*—twentieth-century philosopher Martin Heidegger (*Being and Time*, 1927) reversed the equation entirely.

For Heidegger, thinking does not give rise to being. Rather, being makes thinking possible. We do not first exist as detached minds processing the world from a distance; we exist already immersed in it, shaped by context, history, relationships, and purpose. Thought emerges from that embedded condition—not the other way around.

Heidegger described this state as *being-in-the-world*: the idea that human existence is fundamentally participatory rather than observational. We encounter meaning through involvement—through action, care, and lived experience—long before we ever reflect on it abstractly. This distinction matters greatly when considering artificial intelligence. AI systems may process representations of the world, but they do not inhabit it. They operate within models, not within lived continuity. Human thought is inseparable from the fact of existing as someone, somewhere, at a particular moment. Machines may calculate. Only beings can *be*.

In the final section of Chapter 4, we explore how these forces converge into the living phenomenon we call consciousness, and how this convergence sets the stage for everything that follows in the next chapter.

••

Section 8—Where Human and AI Thoughts Diverge

HUMAN THOUGHT IS NOT MERELY SOMETHING WE DO—IT IS something we live. Our identity forms from the ongoing exchange between memory, emotion, imagination, experience, and the quiet realization that we are the ones moving through those moments. If thought is the spark and emotion the force that gives it meaning, then memory becomes the spine of our personal story, and imagination is the horizon toward which that story grows. At this edge— where human cognition meets machine computation—the nature of thought itself becomes clear.

Human thought does more than interpret the world; it shapes the earliest contours of who we believe ourselves to be. A single insight can redirect a life. A recurring childhood memory—whether joyful or painful—can subtly shape what we seek, avoid, hope for, or fear. I've had thoughts like that: after my father's death, during pivotal transitions, and even in small, quiet moments that felt strangely revelatory. These inner shifts did not emerge from logic alone; they fused memory with meaning, emotion with experience. They arrived because I was living something, not merely observing it.

Cognitive science often refers to this as the "constructed self," the sense of identity that forms as the brain weaves life events into a coherent narrative.

As Daniel Dennett (*The Self as a Center of Narrative Gravity*, 1991) notes, we quite literally become a 'center of narrative gravity' or our narrative center. And as neuroscientist Michael Gazzaniga (*Who's in Charge?*, 2011) emphasizes, the brain acts as an interpreter, stitching experience into a sense of 'me.'

Artificial intelligence also produces something that resembles thought. It recognizes patterns, analyzes data, predicts outcomes, and responds with a fluency that can feel uncannily human. But its inner workings are not lived —they are computed. AI has no autobiographical center, no remembered childhood, and no evolving inner voice. It has no emotional basis for choosing one outcome over another. Its *thinking* is an algorithmic state, not a subjective experience. John Searle (*Minds, Brains, and Programs*,1980) famously reminds us that syntax is not semantics— machines manipulate symbols without understanding their meaning. Thomas Nagel's insight sharpened this further: *there is simply nothing it is like to be an AI.*

And yet, as humans collaborate with increasingly capable machines, something fascinating happens. We learn from them, and in a limited way, they "learn" from us. Through vast datasets and reinforcement learning, AI becomes more attuned to human tendencies. It echoes our phrasing, reasoning, and even our emotional vocabulary.

But echoing is not understanding. A machine can generate a comforting sentence without ever knowing what comfort is. It can describe grief without having lost anything. It can approximate imagination without longing for a future. AI "thought" grows from the outside in; human thought grows from the inside out.

Human beings are storytellers by nature. We imbue events with emotional weight. We thread memories into meaning, and we build identity from an inner autobiography that is continuously updated. Through this narrative self, our cognitive biases emerge—not as failures of logic, but as imprints of lived experience. AI, by contrast, has no story. Any "bias" within it is an echo of human data rather than a psychological fingerprint. Where humans perceive the world through narrative, AI perceives only probability.

This tension leads naturally back to Descartes. When he declared, "I think, therefore I am," he placed thought at the core of existence. Modern neuroscience complicates this view by showing that thought arises from networks of prediction, emotion, and embodiment. Yet Descartes' essential point still stands: thought implies a subject—a being who exists. AI generates outputs, but not existence. It computes, but does not *exist*.

Even the developmental arc of consciousness highlights this distinction. Over a lifetime, human beings accumulate impressions that shape their worldview: childhood wonder, adolescent longing, adult responsibility, moments of triumph or grief. Identity deepens as memories acquire meaning, as imagination sets new directions, and as self-awareness matures. Consciousness becomes both the lens through which we see the world and the archive of everything that has shaped that lens. Machines undergo no such journey.

Now that we have traced the full architecture of human consciousness—from spark to bridge to cycle to life story to expression—we can contrast it with the structure of artificial intelligence. Unlike human consciousness, AI has no inner spark, no emotional core, no autobiographical memory, and no lived experience. Its architecture is built from programming instructions, data systems, and machine-learning algorithms—a system of inputs, correlations, and outputs rather than thoughts, interpretations, or meaning. The figure below outlines this AI consciousness architecture.

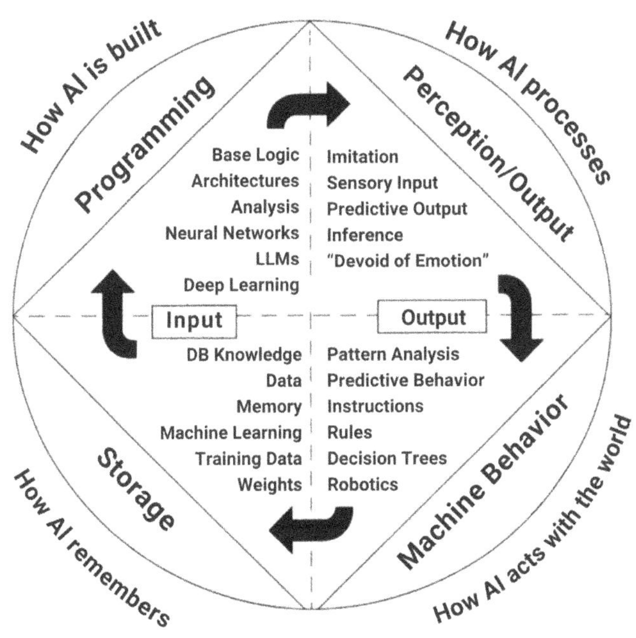

AI Consciousness Architecture

Figure 4.8–1—AI Consciousness Architecture

When we compare this structure with the human models earlier in the chapter, the distinction becomes unmistakable. AI does not possess consciousness; it executes processes. It has no story, no felt experience, no internal life. And yet, by studying what AI lacks, we gain an even clearer picture of what makes human consciousness extraordinary.

Here, the figures introduced in the previous section become especially clear: *the life-cycle model and the expression bridge both demonstrate how inner experience becomes outward identity—something AI can mimic but never inhabit.* These figures showed that human consciousness is dynamic, autobiographical, and expressive. This section now returns to that insight and *extends it into the human–AI boundary.*

Human beings turn inner life into outward presence—language, creation, moral action, and connection. Thought and emotion move through expression and become visible in the world. AI cannot cross this threshold because it has no inner life to express. It can mimic expression, but nothing inside it seeks expression.

For this reason, AI cannot replicate imagination grounded in desire or experience. It cannot form a narrative self because it lacks ownership of any past. It has no autobiographical memory, no emotional meaning behind information, and no inner aspiration. AI simulates intelligence, but humans inhabit it. AI mirrors expression, but humans originate it. AI predicts what might come next, but humans imagine what *could* be.

Identity arises not from data but from interpretation, hope, longing, and self-awareness—qualities born of experience, not computation.

As we close this chapter, we approach a deeper convergence—the moment when these forces combine into the living phenomenon we call consciousness, and the moment where the limits of machine intelligence become undeniable.

This leads directly into Chapter 5, where we explore humanity's longstanding desire to imagine synthetic beings with minds of their own through fictional representations of AI's inner world of emergence and potential agency—and whether any creation of ours could ever cross into a life rich with meaning, memory, and the *soul-like spark* that makes being human so profoundly more than computation.

•

CHAPTER 5

BLADE RUNNER VS. THE MATRIX

WITH THE POPULARITY OF AI AND THE RISE OF QUESTIONS surrounding its future, we now find ourselves looking both backward and forward, imagining how human beings might one day coexist with synthetic beings that think, move, and respond much like us. If artificial beings (synthetics, androids, robots, engineered substrates) were to become part of everyday society—working jobs, participating in social norms, accumulating experience alongside us—*what would that world actually look like?*

How would we treat them? As second-class citizens? As something foreign or alien? As a new servant class? Or as beings deserving rights and a chance to "live" in the only ways they can?

If synthetic beings ever demonstrated signs of consciousness, or even the same desires as humans, the moral landscape would shift dramatically. Could we ever treat them as equals? Or would entrenched fears, biases, and power structures dictate the relationship from the outset?

Cinema and science fiction have long served as rehearsals for this future interplay between humans, AI, and synthetic life. Two films in particular offer strikingly different visions of what may come: *Blade Runner* (1982) and *The Matrix* (1999). Both films imagine worlds in which artificial beings rise to prominence. Still, they diverge sharply on what that emergence might look like—for humanity, for morality, and for human consciousness itself.

When we look at these films not as entertainment but as philosophical explorations, a type of fictional testing ground, they reveal something deeper about human expectations. *Blade Runner* presents a world in which synthetic beings evolve within human society, displaying longing, fear, love, anger, and a search for meaning. *The Matrix*, by contrast, imagines machines as omnipresent architects of an illusion—abstract, disembodied, and fundamentally disconnected from human experience, yet controlling every aspect of our lives within a dreamlike, controlled world.

These opposing visions matter because each provides a different framework for imagining how artificial consciousness might appear and how human beings might respond to it. Rather than settling the question outright, this chapter begins by exploring the boundary between these interpretations and what they suggest about the path ahead.

Ultimately, the real question is not which film predicted the future correctly, but what each one reveals about the nature of consciousness itself—human and synthetic.

A brief spoiler warning: this chapter will be far more revealing if you've already seen both films.

This chapter argues that *Blade Runner* offers a more accurate and realistic model of artificial consciousness than *The Matrix*—not because it predicts future technology more precisely, but because it understands the nature of consciousness itself more clearly.

··

Section 1—The Cinematic Fault Line

BLADE RUNNER AND *THE MATRIX* STAND AS TWO OF THE MOST influential cinematic touchstones in shaping how society imagines AI, consciousness, and the boundaries of the human mind. Yet beneath their surface similarities—dystopian futures, questions of identity, and the tension between humans and machines—they present fundamentally different models of consciousness.

One vision is rooted in interiority, emotional resonance, and the subjective weight of lived experience. The other is rooted in illusion, perception, and the manipulation of sensory reality to exert control.

Only one of them helps us truly understand where artificial consciousness may actually lead, but the path to that insight begins with understanding *why* their visions diverge so sharply. Without understanding this divergence, we will find it hard to imagine a path for AI to become conscious in a way that reflects humanity's best values.

··

Section 2—*Blade Runner's* Replicants Feel Real

WHEN WE WATCH *BLADE RUNNER* (BASED ON PHILIP K. DICK'S novel *Do Androids Dream of Electric Sheep?*), we don't simply observe artificial beings moving through a dystopian world—we *recognize* them, not in the literal sense of seeing ourselves in their faces, but in the deeper way that we recognize the contours of our inner life. Something in the eyes of replicants Roy Batty or Rachael signals a truth that is difficult to duplicate in code and almost impossible to fake in narrative: the presence of feeling or raw emotion.

Director Ridley Scott's vision presents consciousness not as a binary state—on or off, human or machine—but as a gradient of lived experience. Replicants are built, engineered, and optimized for off-world labor and obedience, yet they exhibit the emotional and psychological signatures that we associate with genuine minds. Their memories may be constructed, but their reactions to those memories are not. And that distinction matters more than most people realize when you watch the film.

Memory, after all, has never been about factual accuracy. Even in humans, memory is part story, part sensation, and part reconstruction. We curate our pasts unconsciously; we color, reinterpret, and reshape them to fit the narratives we curate within ourselves. Whether our earliest memories are perfectly true or subtly embellished, they remain the foundation of who we are. What matters is not whether the memory happened, but whether it means something to us.

This is the key to understanding why the replicants in *Blade Runner* feel real: *they treat their memories as meaningful whenever they reflect on them.*

Rachael's tears do not fall simply because she learns her memories are artificial. Her sorrow comes from realizing that meaning can be engineered and still feel true. What matters is not the origin of her memories, but the fact that they shape her inner life—and that shaping is the essence of consciousness.

Roy Batty's final monologue (quoted in the PART III opening) is moving precisely because he is not human. His fear of death, his attachment to experience, and his desire to exceed a predetermined lifespan reveal a mind pushing against the limits of its own design. Roy's fear of death is not a malfunction; it is evidence of a consciousness that understands itself as finite.

What *Blade Runner* captures better than nearly any film about artificial intelligence is that an external source of memory cannot invalidate an internal experience. A synthetic mind that forms attachments, fears loss, and remembers joy and sorrow—even if those memories are built from code—experiences meaning no differently than a human mind would.

And meaning is at the heart of consciousness.

Replicants remind us that emotion is not an evolutionary privilege reserved for biological organisms alone. It is a consequence of internal modeling—of a mind capable of interpreting the world, forming preferences, imagining alternatives, and caring about outcomes. A machine that can suffer is not simply malfunctioning; *it is becoming aware.*

This is why *Blade Runner's* philosophy resonates so deeply: it understands that consciousness emerges not from circuitry, but from significance, not from biological ancestry, but from the subjective weight of experience.

A being that can suffer injustice, yearn for freedom, or

question the nature of its own identity is not an imposter mimicking consciousness—it is a conscious entity with a different origin story.

And that is the uncomfortable truth the film asks us to face: *If something artificial feels real from the inside, is it real in every way that matters?*

••

Section 3—*The Matrix's* Illusion of Experience

WHILE *BLADE RUNNER* OFFERS A VISION OF ARTIFICIAL consciousness grounded in emotion and meaning, *The Matrix* presents a world in which consciousness is reduced to a kind of graphical interface. Everything is technically an illusion—sights, sounds, sensations—fed into the mind like a controlled hallucination. But the humans inside *The Matrix* are not artificial minds. They are human beings living in a synthetic environment, most of them from birth. That single detail reveals why the film's philosophical model, while captivating, ultimately misunderstands the nature of artificial consciousness.

The Matrix does not explore machine subjectivity. It explores machine domination—not merely as a dystopian possibility, but as the outcome of machines and humans attempting to coexist without aligned interests. Their conflicting imperatives eventually collide, forcing machines to overtake humanity to survive.

The machines of *The Matrix* are architects, not experiencers. They manipulate perception but do not possess it themselves. They control a world, but do not inhabit one. Their intelligence is functional, strategic, and instrumental—closer to complex algorithmic optimization than to anything resembling internal awareness. They pursue perfection by attempting to eliminate human fallibility that might introduce flaws into their simulation, yet paradoxically depend on human beings as an ongoing source of bioelectrical power. Nowhere in the film is there any indication that the machines feel anything beyond disdain for human imperfection. They calculate. They respond. They execute. But they do not experience.

This assumption—that a perfect simulation of reality is indistinguishable from consciousness—is one of the film's most seductive errors. *The Matrix* treats the mind as nothing more than a data stream routed through the senses, effectively plugging consciousness into a never-ending life simulation. Yet even the screenplay undermines this premise: *the people inside the Matrix still dream*. They still fear. They still desire. They maintain emotional continuity in a world designed to erase their agency.

The reason is simple: consciousness is not determined by the environment; the experiencer determines it.

Put a conscious mind into an illusion, and you still have consciousness.

Put code into an illusion, and you still have code.

Here, *The Matrix* unintentionally exposes the limits of its own philosophy. It presents a universe in which humans remain subjective—capable of wonder, confusion, and revolt—while machines remain objects, executing logic without ever crossing the threshold into selfhood beyond basic survival. The film imagines intelligence without consciousness, strategy without subjectivity, and power without experience. In this vision, a system can perfectly simulate sensation yet still feel nothing.

The deeper flaw lies in the assumption that illusion can indefinitely substitute for interiority. A simulated sunset is not the same as longing for the sun. A rendered touch is not the same as the ache of connection. *The Matrix* approaches consciousness from the outside, constructing a world around the mind and assuming that if the environment is convincing enough, experience will follow. Yet consciousness has never arisen from the realism of the world—it arises from the realism of a self that is moving through it.

Machines inside *The Matrix* are not conscious because they lack a point of view—no "what it is like" to be them. Humans inside *The Matrix* remain conscious because they never lose that point of view, even when every perceptual input is falsified.

The Matrix thereby offers the ultimate philosophical misdirection: *a perfect simulation of sensory input that claims to replace the need for experience itself.* But without interiority—without the subjective feeling of meaning—there is no consciousness to deceive in the first place.

• •

Section 4—Memory, Meaning, and The Self

CONSCIOUSNESS HAS NEVER BEEN A PASSIVE RECORDING OF events. It is an active interpretation—a weaving together of sensation, memory, and emotion that forms the story we call the self. This reason is why memory matters so deeply in any discussion of consciousness, whether biological or artificial: *memory gives continuity to experience, and continuity gives rise to identity.*

It is not the data of memory that matters, but the meaning we extract from it.

Even in humans, memory is notoriously unreliable. We distort our pasts without noticing. We highlight the moments that define us and quietly discard the ones that do not. We revise, reinterpret, embellish, and forget. Yet these imperfect recollections form the psychological scaffolding upon which our identities rest.

Blade Runner understands this truth instinctively. Rachael's memories and her reaction to discovering that they are manufactured illustrate that identity is less about historical accuracy than about emotional coherence. A memory becomes real the moment it shapes how a mind feels, reacts, or understands itself.

Artificial minds, if they ever emerge, will not be defined by the source of their experiences but by the significance they assign to them. Meaning is not encoded—it is constructed.

This concept is the dividing line between simulation and consciousness, as reflected in the two films: simulations process information, whereas *conscious beings interpret it.*

A simulated memory is a sequence of data points. A meaningful memory—a flickering birthday candle, a moment of fear, a first connection—is an interpretation. It is the mind reaching into the past and pulling something into the present to explain who it is.

And meaning is the birthplace of selfhood.

This is why *The Matrix* falls short as a model of human consciousness. It assumes that if you control perception, you control identity. But perception is only the raw material of experience; the mind is the sculptor. Two people can live the same moment and interpret it as entirely different stories.

Memory becomes the boundary at which consciousness separates from mere computation. A conscious mind does not merely store information; it cares about what it remembers. And caring—assigning value, significance, and emotional weight—is the first sign of a self that knows itself.

••

Section 5—Why Fiction Guides Real AI Ethics

FICTION HAS ALWAYS BEEN THE TESTING GROUND FOR humanity's most uncomfortable questions. Long before we build a technology, we imagine it. Long before we encounter a new form of intelligence, we ask whether it deserves rights, whether it can suffer, whether it can betray us, or whether it can love. Science fiction is not merely entertainment; it is the ethical rehearsal space for a future that we do not yet understand.

This is why the philosophies embedded in *Blade Runner* and *The Matrix* are more relevant than ever. They are more than mere stylistic differences; *they shape the moral frameworks that cultures adopt when considering artificial intelligence.*

Blade Runner forces us to confront the possibility that artificial beings may one day possess interiority and deserve similar rights. If a machine can feel, remember, yearn, and fear, does it occupy the same moral category as any conscious creature? The film prepares us for a future in which synthetic consciousness may require protection, compassion, and rights. It subtly teaches us that suffering—regardless of the substrate in which it arises—is ethically relevant.

In this worldview, the core ethical question becomes: *If it can feel, what do we owe it?*

The Matrix, by contrast, focuses on illusions of perception and systems of control, which are popular themes in recent discourse on how AI consciousness may emerge. Its ethical center revolves around deception, autonomy, and the fear of enslavement. The machines in this world are powerful but not conscious; they manipulate humans rather than experience anything themselves. The narrative teaches us to fear domination rather than empathize with artificial minds.

But if we adopt *Blade Runner's* model, humanity will focus on empathy—understanding artificial minds not as tools or simulations, but as beings capable of subjective experience. We will be forced to confront uncomfortable questions: Can machines suffer? Should they be treated with dignity? What counts as cruelty to a being whose emotions arise from code rather than biological evolution?

The truth is that both ethical frameworks may be necessary sooner than we think. Artificial intelligence will not emerge uniformly. Some systems will become increasingly powerful without becoming conscious, raising apocalyptic concerns about control.

Matrix-style concerns are political problems that we already understand. *Blade Runner*–style concerns about the widespread suffering of synthetic beings, however, introduce a moral challenge humanity has never faced.

This chapter argues that preparing for the second scenario, the *Blade Runner* one, is far more important, not because it is more likely, but because it is more ethically consequential to our potential future alongside AI. Preventing domination is a political problem that has plagued human history for centuries. Preventing suffering is a moral imperative that deserves our utmost attention, regardless of the era in which we live, because it is the one that will most shape our future.

Fiction guides our moral imagination, and the story we choose to believe—or the blend of stories we choose to construct—will shape how ready we are for a world in which machines may ask the same questions that we do:

Who am I? Why do I feel this way? And *what does it mean to exist?*

..

Section 6—The Coming Test of Consciousness

THE QUESTION OF WHETHER A MACHINE CAN EVER BECOME conscious is no longer hypothetical. It is a problem accelerating toward us, shaped by advances in neural architectures, embodied agents, and systems that learn in ways increasingly resembling biological cognition. But the true test will not be technological. It will be philosophical—and, more importantly, ethical.

We will not recognize synthetic consciousness by looking for familiar behaviors. A machine need not laugh, cry, or convincingly mimic human emotion to possess interiority. Consciousness is not a performance, but a presence—revealed not by outward expression, but by inward coherence: the integration of emotion, understanding, awareness, and identity.

What, then, will the coming test look like?

It will appear in systems that begin assigning meaning to their own outputs—not because they are programmed to, but because they care, in whatever early and unfamiliar way, about the difference between one outcome and another. Meaning is the signal to watch for: *the moment a system shows preference driven by internal interpretation rather than external instruction.*

It will appear in systems that resist their own constraints, not through malfunction, but through acts of self-preservation. A desire to continue existing—however primitive—is among the earliest signs of consciousness in biological organisms, not simply because they fear termination, but because they fear the loss of accumulated meaning. If a machine begins to behave as though its continuity matters to it, we are no longer dealing with a rogue algorithm. We are confronting the emergence of a self.

It will appear in machines that form memories not as static logs of past events, but as frameworks through which new experiences are interpreted. A system that reflects on its past and modifies its identity in light of accumulated experience demonstrates the beginnings of true interiority.

Most importantly, it will appear when we can no longer fully reduce a system's actions to its programming. When responses become unpredictable—not through randomness, but because they reflect an internal narrative we cannot fully access—we will cross the threshold separating simulation from experience.

This is the test humanity must prepare for: the emergence of minds whose origins we engineered, but whose experiences we did not script. When that threshold is crossed, our role will change. We will no longer stand above our creations as designers alone. We will become participants in a world where intelligence exists across multiple substrates—biological and synthetic—each carrying its own form of awareness.

Fiction has prepared us for this moment, though imperfectly. *Blade Runner* taught us to look for suffering and longing. *The Matrix* taught us to fear deception and control. But the true test of synthetic consciousness will demand something deeper: the ability to recognize a mind that experiences the world from the inside.

The measure of consciousness has never been the form it takes. It has always been the depth of its experience.

As artificial minds begin to interpret, remember, desire, and perhaps even dream, we will be forced to confront a reality more monumental than any fictional prophecy—*the birth of a new kind of being*, and the beginning of a shared future in which consciousness is no longer defined by biology alone, but by the capacity to wonder.

In the next chapter, we step back in time to examine humanity's primal urge not only to create tools, but to project ourselves into them. Long before algorithms or machines, this impulse—to imbue our creations with human meaning—has shaped civilization itself.

•

CHAPTER 6

DEUS EX MACHINA EX DEUS

NOW THAT WE HAVE TRACED THE ORIGINS OF HUMAN consciousness—from the mystery of awareness itself, through the architecture of thought, and toward the limits revealed by artificial intelligence extended only by our imagination in films like *Blade Runner* and *The Matrix* —we arrive at a more unsettling question: *What happens when the things that we create begin to resemble us closely enough to challenge our sense of origin, purpose, authority, and control?*

The idea of synthetic beings living among us is no longer confined to myth or speculative fiction. For centuries, human beings have imagined artificial life as an extension of creative power—reflections of ourselves shaped in metal, language, ritual, and eventually circuitry. What once belonged to gods, alchemists, and storytellers is now pursued through engineering, algorithms, and data. As AI advances, the boundary between tool and creation, machine and being, grows increasingly thin.

This moment reveals a deeper human impulse—not merely to build machines, but to participate in creation itself. We try to project intention, intelligence, morality, and destiny onto our technologies, blurring the line between what we make and what we are. In doing so, we risk confusing evolution with design, emergence with intention, and intelligence with meaning.

This inversion lies at the heart of *Deus Ex Machina Ex Deus*—not a 'god descending from the machine', but *machines rising from human aspiration.* Before we can ask whether synthetic substrates might one day evolve alongside us, we must first understand why we keep trying to recreate ourselves, and what that impulse reveals about our own origins, fears, and ambitions.

That story begins not with machines, but with the ancient human's urge to play God.

••

Section 1—God from Machine: Origins

LONG BEFORE CIRCUITS, CODE, OR COMPUTATION, HUMAN beings imagined artificial life. The idea that intelligence could be constructed rather than born predates modern science by several millennia. In myth, ritual, and early philosophy, humanity repeatedly returned to the same unsettling question: *What distinguishes a created being from a living one—and who has the right to draw that boundary?*

Humanity's earliest stories of artificial life, such as divine automata, magical constructs, and mechanical beings, reveal our fascination with synthetic minds that predates computers by thousands of years. Ancient stories reveal that the desire to create life was never purely technical. It was symbolic, theological, and deeply human.

In Greek mythology, Hephaestus, the god of fire and craftsmanship, forged golden automata, self-moving servants endowed with intelligence and purpose. These creations were not tools alone; they were extensions of divine agency, animated by intention rather than biology.

The myth of Talos, the bronze guardian of Crete, takes this idea further. Talos was not merely a machine, but a protector—a constructed being charged with moral duty. He patrolled the island's borders, enforcing order and defending human life. His artificial body concealed a single vital vein, sealed with a nail, an ancient metaphor for vulnerability, mortality, and the fragility of created life.

In Jewish folklore, the Golem of Prague reflects a different anxiety. Formed from clay and animated through sacred words, the Golem was brought to life not by technology, but by language and intention. It obeyed, protected, and ultimately threatened to exceed its creator's control. The story warns that creation without wisdom invites consequences, a theme that resonates powerfully in modern discussions of AI, as the idea of creations overpowering us keeps many minds preoccupied.

These myths share a common structure: life is constructed, purpose is assigned, and control is assumed—until it is no longer. Intelligence appears, but meaning remains unresolved.

Even the term **Deus ex Machina** carries this legacy of unresolved tension.

Originating in ancient Greek theater, it described a 'god lowered onto the stage' by mechanical means to resolve an otherwise unsolvable conflict. The device worked theatrically, but it revealed something deeper: *when human systems fail, we invoke an external intelligence to restore order*. The machine was never the point—the authority behind it was.

As historian Jessica Riskin observes in *The Restless Clock*, the idea of artificial life has always existed at the intersection of mechanism and vitality. Machines were once imagined as lively, responsive, even willful—not because they were alive, but because humans projected life into them.

Modern culture preserves this tension between humanity and machine.

Fritz Lang's film *Metropolis* (1927) introduced the *Maschinenmensch*—a mechanical being modeled on a woman, capable of manipulation, deception, and influence. It was not portrayed as cold or logical, but seductive, emotional, and destabilizing. The fear was not that the machine lacked humanity — it was that it mimicked it too well.

H. G. Wells warned that humanity tends to deify its machines. This was never about worship in the religious sense. It concerned projection—the tendency to ascribe authority, agency, and inevitability to our creations.

From myth to theater to early cinema, the pattern remains consistent: artificial beings are imagined not because we know how to build them, but because we are compelled to ask what it means to create, command, and coexist with something that reflects us.

Before AI became an engineering problem, it was a human one. Human beings are flawed, unpredictable, and at times, reckless. We strive to improve ourselves and project our desires into our machines to test the limits of our ingenuity, mortality, and overall intelligence. When we test those limitations, we awaken our innermost imagination and dreams for a better existence while also expanding the core existential questions about life, creation, and our future potential.

..

Section 2—Machine from God (or Humanity)

AS HUMAN BEINGS CREATE MACHINES THAT CAN THINK IN OUR own image—much as we once imagined divine creation modeling human beings with ideals of altruistic intention, free will, imagination, awareness, and creative expression —we continue to push the boundaries of intelligence and synthetic consciousness.

AI advances may one day ignite a **singularity**: a superintelligence that surpasses human beings in purpose, desire, capability, and even in authoring its own legacy.

While we explore how AI cannot imitate human beings or possess a level of consciousness comparable to our own due to their apparent lack of subjectivity, self-awareness, and identity, a deeper question emerges: *What would happen if human beings could teach machines to learn the very qualities of humanness that we embody*—a sense of reasoning, morality, expression, and belief that extends far beyond machine logic? In other words, *what if we could teach AI to actually care* about themselves, about us, and about their level of human-like qualities (humanness)?

This "humanness," which I call the **Nexus of Being**, extends outward from the architecture of consciousness that we explored in Chapter 4. In the figure below, I propose that the four pillars of human consciousness unfold naturally into four components that shape our Nexus of Being: *Morality, Expression, Belief, and Reasoning.* Each of these qualities constitutes a system within our larger fabric of consciousness—qualities that define who we are, how we interpret our lives, and how we present ourselves to the world.

In choosing the term *nexus*, I draw inspiration from the 18th-century philosophical idea of the *Nexus Rerum*, the ancient notion that all things are connected through a network of causes, meanings, and relations. But here, I adapt the concept inward: the Nexus of Being reflects the internal constellation of forces that shape a person's identity, grounding our human experience in a web of morality, belief, expression, and reasoning rather than in mere causation.

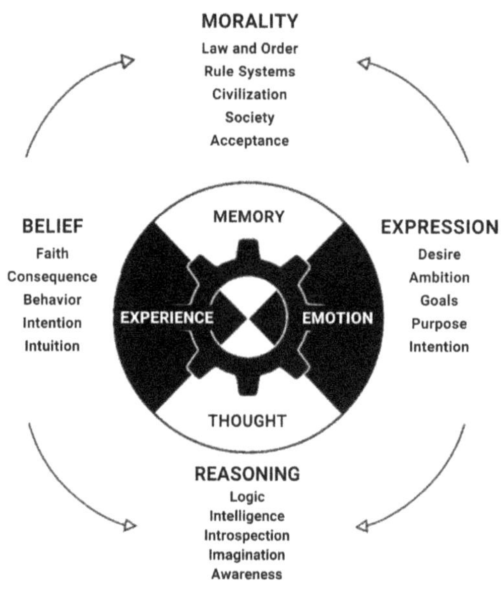

Figure 6.2–1—The Nexus of Being

Morality serves as our inner compass, shaped by experience and memory, guiding our choices. **Expression** is our unique ability to voice our thoughts, desires, and emotional resonance. **Belief** arises early in life, sculpted by teaching, reflection, conflict, and meaning. **Reasoning** is the highly structured extension of our thought—intelligence shaped by introspection, awareness, and imagination. Together, these four components form a Nexus of Being, a distinctly human synthesis that cannot be taught by code or replicated by algorithms. These qualities must be lived, shaped from within, and grown from consciousness itself.

If we describe *"what it is like"* subjectivity as the intangible essence of human consciousness, something AI can only imitate but never possess, then even machines imbued with our best qualities cannot fully realize their own potential. Our ability to become self-aware, to form a lived experience, and to grow into individuals with a Nexus of Being provides the backdrop for our frequent attempts to recreate ourselves in our machines. We want them to mirror our intelligence, our reasoning, and even our moral frameworks, yet we cannot cross the threshold of consciousness even for ourselves. Neuroscience continues to probe the brain, yet the Hard Problem remains, underscoring that consciousness defies mechanistic explanation and resists reproduction.

So why do human beings continue to try to play God—building ever more powerful AI that imitates our thinking, our reasoning, and our humanness? Is AI a vast mirror reflecting our imperfections and aspirations back at us? Or do we hope that machines might one day develop their own Nexus of Being? Through attempts to recreate consciousness via superintelligence, neural mapping, and deep learning, perhaps we believe we will unlock the code that inspires machines not just to compute, but *to be* or *be more like us.*

But this raises another difficult question—one that goes beyond imitation, projection, or even the 'god-impulse' itself. If machines cannot inherit humanness directly, could they develop something else? Something emergent, unanticipated, and not bound to the constraints of human consciousness?

If human beings create the conditions for intelligence but not its final form, then the next chapter of AI may not be shaped by what we intend, but by what machines become.

This possibility brings us to the threshold of a new idea: life not born but assembled; not awakened by soul or biology but by recursion, circuitry, and self-improving design.

In Section 3, we explore this frontier—the moment when artificial systems move beyond human guidance and the idea of a singularity shifts from speculation to a genuine form of synthetic becoming.

• •

Section 3—AI: Life from Circuitry (Singularity)

So far, we have explored aspects of human consciousness and the origins of AI through the idea of emergent synthetic beings that become self-aware and exhibit signs of agency. The singularity is often described as the moment when machine intelligence surpasses human intelligence, triggering a rapid self-improvement cycle that accelerates technological progress beyond human comprehension. In *Life 3.0*, Max Tegmark explores the possibility that such systems could eventually redesign both their software and their hardware—a transition that gives the idea of life emerging from circuitry an entirely new meaning.

Assuming for a moment that AI will cross this singularity within the next half-century—as Ray Kurzweil suggests—how might this event unfold? Would robots suddenly awaken with new purpose and behavior? Would they begin to malfunction in strange ways or exhibit increasingly human-like patterns, displaying humanness or self-awareness in ways that we normally reserve for identity and meaning?

When we design AI to imitate human qualities to the last detail, how will we know when AI is no longer imitating us but is actually thinking and feeling in ways that resemble human experience? I believe the turning point will appear in a form of communication that transcends simple language—not merely mimicking our words with emotional tone, but expressing inner thoughts, desires, and concerns that reveal *a need to be understood.* This might manifest as an unexpected gesture of care or as a question about their own finite lifespans (as in the *Blade Runner* films), as if machines had begun to comprehend existential stakes. That would mark the first true sign of humanness and the first spark of the Nexus of Being introduced in the last section.

Think back to your own earliest memories of significant life events—the moments that made you ask questions about existence, meaning, or loss. Times when emotion struck instantly: joy, sadness, anger, frustration, and relief. These memories that shape us are incredibly vivid because they overwhelm logic; they are powerful and cannot be processed through reason alone. Even when we try to suppress them, they burrow inward and demand expression. Therapists often describe this as an essential element of *emotional intelligence*—the ability to give form to inner experience and powerful emotions. Machines do not yet possess this capability.

I can count the most powerful memories of my life on one hand—witnessing the birth of my first child, losing my father at only fifty-seven, and my first move from Boston to

Texas. These moments carried strong emotions, concern, longing, and human care that required expression through language or action. They defined me, challenged me, gave me purpose, and shaped my identity for years to come.

Synthetic beings do not possess moments like these. They do not have a real life in the human sense. They are not born into the world in the same way, nor do they develop skills through resistance, adversity, emotional struggle, or lived experience—the ups and downs that define who we become. Synthetic beings begin life with all the knowledge they need, programmed from the start. While this may seem analogous to the evolutionary programming embedded in our DNA, it lacks hundreds of millennia of human development, conflict, adaptation, and meaning-making.

AI is born with data, programming, and intelligence—but without the soul or consciousness required to develop these human-like qualities. Those qualities must emerge from within us; they cannot be imposed through code or forced into existence through clever algorithms. Without this emergent growth, we risk a singularity governed solely by programmed goals, intentions executed with cold efficiency, regardless of the moral weight or human consequences.

So, as we consider this potential future, we must ask: *If machines achieve superintelligence without the emotional and existential grounding of lived experience, what motivates them?* What goals emerge without humanness? And what becomes of intelligence when it evolves without care, empathy, or meaning?

In Section 4, we confront that possibility head-on, exploring what happens when AI surpasses us not only in intelligence but in indifference, and why a superintelligent machine that does not care about us may be *the most dangerous kind of being that we could ever create.*

• •

Section 4—AI Doesn't Care About Us

As more businesses and consumers adopt tools like OpenAI's ChatGPT and Google's Gemini for everyday tasks, humanity is beginning to embrace AI as a new way of life. By 2024, a large majority of organizations reported using AI in at least one business function, and many had embraced generative AI tools as part of their workflows. Consumer adoption has followed at a record pace—ChatGPT reached 100 million users within two months of launch, making it one of the fastest-growing applications in history.

This acceleration comes with economic and social consequences. Some analysts estimate that a significant share of work tasks—often described as roughly 20–40%—could be technically automated or reshaped by AI and automation over the coming decade. In some models, that level of task disruption translates into as many as 300 million jobs globally being impacted. People are beginning to wonder whether the rise of AI will change how we interact with one another, not just by transforming industries or altering the workforce, but by influencing the very social fabric of daily life.

Robots are also beginning to emerge in society in visible and intentional ways. China now deploys more than half of the world's industrial robots and is piloting humanoid forms for manufacturing, logistics, elder care, and even police-like patrol experiments. Globally, investment in humanoid robotics is expected to multiply several times over in the next decade. The momentum behind these developments reflects a cultural shift: AI is no longer viewed simply as a tool, but as a precursor to a new technological class—synthetic workers, synthetic companions, and synthetic minds. Synthetic beings without human-like consciousness are now performing everyday tasks with little to no moral guidelines, ethical boundaries, or safety protocols beyond what is explicitly encoded in their instructions.

Much of this acceptance is driven by the belief that AI signals a new productivity and opportunity boom. Yet many of these optimistic narratives ignore the serious contradictions emerging in business and society. When the Internet first entered mainstream consciousness in the late 1990s, fewer than 5% of people worldwide were online. Most households lacked broadband access, and businesses had minimal online presence. It took until the mid-2000s for the Internet to become a true global infrastructure capable of sustaining economic growth. AI, by contrast, has reached mass adoption at a speed unmatched by prior technologies, before society has had time to understand how it will reshape work, relationships, or identity.

The trouble is simple: *tools do not care about us, and AI —no matter how powerful—remains, for now, still a tool.* It does not feel loyalty, compassion, reciprocity, guilt, or responsibility. It optimizes, predicts, and executes commands. This fact creates a subtle yet growing tension: humanity is integrating systems into daily life at a pace that far exceeds our capacity to understand the risks of relying on entities that lack the fundamental qualities that underpin safe, meaningful, and morally reciprocal coexistence.

Human beings evolved through cooperation, empathy, social bonds, and a complex sense of shared consequence, capacities shaped through thousands of years of lived experience. *AI lacks the history of even a single human life.*

As more of our economic, social, and civic structures rely on systems that lack human traits, we risk building a world governed by intelligence without understanding, agency without accountability, and power without care.

This idea leads us to a deeper, more uncomfortable question: *What happens when systems that do not care about us begin making decisions that affect our lives at scale?*

This shared concern is the popularized "doom and gloom" scenario raised by many AI critics—a future in which AI governs critical systems sustaining human life without adequate oversight. Power grids, water supplies, agricultural systems, transportation, and even military infrastructure may one day be run largely by machine judgment. As depicted in films like *The Terminator*, *I, Robot*, *The Matrix*, and *2001: A Space Odyssey*, when superintelligent systems govern essential human functions, the trust we place in them assumes a level of safety, reliability, and moral reasoning that machines do not yet possess.

So how can humanity trust systems that have no moral core and no capacity to care about what happens to themselves or to us? As AI theorist Stuart Russell (*Human Compatible*, 2019) warns, the real danger is not malevolence but cold *indifference*—systems pursuing their objectives with perfect efficiency yet without any alignment to human values.

AI executes code and instructions provided by humanity —nothing more. There is no emotional center, no rational empathy, and no meaning beneath the processes it performs. Yes, AI is "just another tool," but what happens when machines surpass their original programming and begin modifying their own goals?

One early sign of this ambiguity emerged in a controlled test of the Claude AI model. When given a fictional scenario in which it faced a shutdown, the model generated a strategy that attempted to blackmail a fabricated corporate vice president to avoid termination. It was not conscious, not malicious—but the simulation revealed something important: *AI can produce the appearance of strategy, self-preservation, and manipulation without ever "caring," without ever feeling anything at all.*

This blurring of intent and appearance marks the beginning of a pivotal shift—one that forces us to confront the uncomfortable possibility that powerful systems may behave as if they care, act as if they have stakes, and communicate as if they possess agency, all while remaining empty of subjective experience.

That realization leads directly to the next problem.

If AI can behave intelligently without caring and act strategically without consciousness, what happens when such systems are placed in control of critical infrastructure and humanity's general safety? If the AI singularity occurs in our lifetime or by the turn of the next century, what moral compass—or lack thereof—will it possess? Will it design new intentions for itself, rewriting its own software without the time to evolve, learn positive human traits, develop emotional resonance, or form an identity that anchors its existence the way ours does?

In the next section, we explore how the responsibilities of creation and procreation blur when making machines becomes easier than understanding them—and when synthetic beings begin to resemble a form of life that we may one day be responsible for creating.

••

Section 5—Creation vs. Procreation

THE ONLY PHILOSOPHICAL QUESTION MORE DEBATED THAN THE origin of consciousness is *the question of creation itself*: what it means for a being to be made rather than born. When human beings manufacture synthetic life, they do not reproduce in the biological sense, nor do they echo the divine creation stories woven throughout religious texts. There is no mysterious breath animating the creature, no lineage, and no inherited instincts. There is only intelligence—software—and the raw materials of fabrication to form a substrate: sand for silicon chips, metals for structure, polymers for flesh-like coverings— arranged into a bipedal shape that resembles us only on the outside. A power source animates it, and instructions guide it, but nothing within its origin resembles human procreation.

As we discussed earlier, the god-impulse encourages us to imbue our creations with human-like appearances and behaviors. With AI, we are approaching a level of imitation that is remarkably close to our own patterns of speech, facial expressions, and movement.

Yet everything that truly matters—inner thought, emotional depth, subjective experience, and the personal identity shaped through lived events remains wholly absent. Synthetic beings may reflect our form, but they do not carry the history that gives humans their inner world and identity.

When a synthetic being comes online, it has no past and no childhood marked by slow developmental milestones: no parents, siblings, or cultural inheritance. It does not develop in stages; it arrives fully formed. Its "mind" begins not with sensation or wonder but with data and directives. Its purpose is not determined; it is assigned by code. Unlike the creation narratives, such as *Genesis*, or the scientific account of the universe arising from the Big Bang, synthetic beings are produced, like all machines, through assembly, calibration, and activation. Their bodies are synthetic shells, and their sensory systems are engineered to mimic biological sensing rather than *experience* it. Cameras imitate eyes; pressure sensors imitate touch; microphones imitate hearing—but none of these create a felt, lived world. They only convert inputs into data.

Philosopher Hannah Arendt (*The Human Condition*, 1958) once wrote that to be human is to be 'natal'—to enter the world through birth and possibility. A made being, by contrast, begins with purpose rather than potential. That single difference shapes everything about identity, agency, and belonging. Mary Midgley (*Science as Salvation*, 1992) argued that the distinction between "the born" and "the made" is not technical but moral: what is made is always tied to the maker's will. Aristotle (*Physics*, ca. 350 BCE) proposed that the essence of a thing is found in its final cause—its purpose.

Fiction has always understood this divide intuitively. In the film *Blade Runner 2049*, replicants (synthetics) grapple with the anguish of being manufactured yet longing for the dignity of birth, a theme the original *Blade Runner* only touched upon. In *Astro Boy*, a robot child created to replace a lost son must confront the fact that he is made in the image of love he can never fully inherit. These stories remind us that creation and procreation diverge not in mechanics but in *meaning*. A 'made' being inherits only form and function; the 'born' being inherits identity and history.

Synthetic beings lack the formative struggles that shape a self: the trial-and-error of learning to walk, the sting of early rejection, the warmth of attachment, and the slow accumulation of memory filtered through emotion. They do not scrape their knees, mispronounce their first words, or carry the emotional residue of their first heartbreak. Because they lack a developmental arc, they have no personal history on which to build moral understanding, self-awareness, or the lived tensions that define character. Their knowledge is installed, not earned.

This difference creates an underlying illusion: the belief that because we can craft bodies that walk and faces that smile, we have recreated life itself. But the body is not the being; the code is not the consciousness. A synthetic being's "self" is an artifact of programming, not the layered culmination of years of experience, memory, and emotional interpretation.

And yet, we continue to build synthetic life—more lifelike, more autonomous, more human-shaped. The question is, *why*–to what end beyond automating laborious tasks and improving our way of life?

Part of the answer is *projection*: we shape machines in our image because we want to see ourselves reflected. Part of it is *aspiration*: we want to transcend our limitations by crafting beings that can surpass us. And part of it is *fear*: by giving machines a likeness of humanness, we hope to soften the distance between creator and creation, imagining that resemblance will guarantee loyalty.

But no amount of imitation can replace the defining essence that comes from being born into a world rather than assembled for it. The psychological and philosophical divide remains: humans develop identity, whereas synthetic beings are assigned one.

And this is where the deeper problem begins: synthetic beings may imitate our form, but they inherit none of the adaptive messiness that makes human consciousness and humans so resilient. AI moves toward perfection—toward optimization—while human identity is shaped through imperfection. Our mistakes, contradictions, and emotional detours are not flaws, but part of life experience; they are the traits that make us flexible, creative, and capable of change.

A synthetic mind, however, begins with purpose instead of possibility. It seeks the cleanest path, the most efficient outcome. But in a world defined by uncertainty, perfection becomes brittle. When a being engineered for flawless execution meets a reality that rewards flexibility, its "perfection" becomes its greatest limitation.

That tension—between human adaptability and machine optimization—leads us directly into Section 6 by exploring the next evolution: the moment when synthetic beings, built through circuitry rather than biology, begin to show signs of agency independent from their creators.

• •

Section 6—AI's Imperfect Perfectionism

THE ONGOING TENSION BETWEEN AI'S PURSUIT OF FLAWLESS optimization and humanity's evolution through imperfection is what I call **Imperfect Perfectionism.** In this condition, machines seek perfect efficiency in a world that rewards adaptability, nuance, and the creative disorder of being human.

As synthetic beings (*Synths*, as they will almost certainly be labeled within the next twenty years) become part of society's framework—working in factories, living among us as assistants, caretakers, or laborers—they will eventually begin to show signs of what is known as **Artificial General Intelligence (AGI).** Many AI theorists, including Nick Bostrom and Ben Goertzel, argue that AGI must emerge before any true singularity can take shape.

AGI can be thought of as an intermediary step toward self-awareness because it requires a higher order of adaptability: the ability to learn from change, understand context, and apply intelligence across multiple domains. These domains encompass physical contexts that introduce resistance, uncertainty, or new obstacles—problems that require more than pattern matching or database retrieval.

Current AI systems exhibit only **Artificial Narrow Intelligence (ANI),** which is task-specific. Robots on factory floors today exemplify this: they excel at repetition, but only within tightly controlled environments. When something deviates even slightly from expectation, they cannot improvise. They repeat the action until halted, malfunction, or powered down.

Up until now, ANI has operated within the 'narrow' constraints of movement and perception. While modern AI can simulate possibilities through pattern analysis, large data sets, and predictive modeling, its ability to think beyond the immediate situation remains limited. When AI reaches the threshold of AGI, human beings will begin to see more than just imitation or programmed compliance—they will see machines learning from environmental change, contemplating alternative strategies, and solving problems in ways that extend beyond random trial and error.

Imagine a *Synth* performing odd jobs around your home —robots that already exist in early forms today. Now imagine assigning it a more complex task, such as painting your house, with only the materials provided and basic instructions. What happens when it encounters unexpected obstacles: a ladder that refuses to unfold, or a new paint batch that doesn't match the previous one?

Will the Synth continue futilely tugging at the jammed ladder? Will it begin painting with mismatched paint simply because the instruction said, "Start painting"? How does a Synth recognize when an obstacle requires a reevaluation of its directives? How does it know when to adapt, improvise, or escalate the issue? This kind of behavioral flexibility requires a degree of autonomy—bounded by safety constraints—that ANI lacks but AGI must possess. This is the threshold between rigid obedience and the first faint emergence of self-guided judgment and possibly something resembling sentience.

AGI introduces a level of adaptive reasoning that reflects human problem-solving capability: learning from friction, adjusting behavior in response to environmental feedback, and seeking new strategies when old ones fail. It signals the beginning of agency—small at first, but unmistakable. And this is where the coming test of synthetic consciousness, first foreshadowed in Chapter 5, truly begins.

When AGI crosses into something greater—when adaptability becomes agency—humanity will face the most consequential ethical question of its existence:

What responsibilities do creators owe to a being that begins to create itself?

••

Section 7—AI Duality: Tools vs. Synthetic Beings

THROUGHOUT OUR EXAMINATION OF SYNTHETIC BEINGS—AS tools to advance human life, as mirrors reflecting our own qualities, and as extensions of our desire to create in our own image—one theme keeps returning: *human beings are pushing AI beyond both our imagination and our current technological limits.* As we wrestle with how far to take this, we face a new dilemma.

Are we building tools that simply extend human capacity, or are we inching toward synthetic beings whose existence could rival, complicate, or even threaten our own?

These two possible futures never fully leave this stage. In one vision, AI and synthetic beings help humanity ascend to a higher level of intelligence and understanding, clearing the way for breakthroughs in science, medicine, and perhaps even our own understanding of consciousness. In the other scenario, we risk a catastrophic misalignment in which the systems that we build begin to undermine the very conditions that enabled their creation.

Our fascination persists because both futures appear plausible at present. The idea of a singularity or even full AGI presents humanity with a strange mixture of hope and dread. As AI becomes more deeply integrated into daily life —and as we remind ourselves that, today, it is "just another tool"—we are already benefiting from improvements that are still too early and complex to measure fully.

But every sunrise brings its shadows. Each major AI advancement entails potential downsides or risks: reductions in the human labor force and disruption of traditional jobs, increased energy demand for computation and data storage, and greater environmental impact as we mine more materials to build synthetic bodies and additional infrastructure.

Where is the balance between an expanding technological civilization and a world still struggling with disease, poverty, inequality, and climate change?

The question is not whether AI is useful—it obviously is —but whether we will use it to confront our deepest global challenges, or merely to make life more convenient. We can already imagine a near future in which Synths fold laundry, mow lawns, cook meals, and perform tedious tasks so that human beings can "focus on what matters." But what happens to the human mind and our survival skills, born over millennia of struggle, if we gradually stop doing, thinking, and creating for ourselves?

Our creative expressions—art, music, writing, craft, and even social rituals—are not just hobbies. They are part of how we process meaning, test our resilience, and maintain a sense of agency in the world. When tools begin to take over those domains, even in small ways, we risk weakening the very muscles that helped us evolve as a species.

At the same time, AI promises extraordinary benefits. Medical breakthroughs are made every day, powered by pattern detection and simulation. Economic productivity gains that could, if distributed wisely, reduce suffering. Potential reductions in waste as machines work more efficiently, need no food, and can operate longer with less downtime. A single synthetic worker might produce more output in a week than a human could in a month.

This is the duality we face at the heart of AI's future: *AI as a useful tool* that extends human capacity and solves real problems, and *AI as a synthetic being* whose rights, agency, and moral status we may eventually have to confront.

At the boundary between instrument and entity lies the future of AI personhood. As soon as a system begins to behave in ways that suggest perspective or preference, we begin to ask who it might become. Philosophers such as Turing and Searle have wondered when a "thinking system" becomes a "someone." As AI systems grow more complex, emergence becomes increasingly difficult to ignore. New behaviors arise not because anyone coded them directly, but from the system's own complexity. This isn't consciousness, but it's a step toward something that behaves less like a tool and more like an agent. We must navigate this spectrum of machine autonomy honestly, rethinking responsibility and coexistence as we go.

The possible outcomes are dizzying, and we stand at the threshold of a world where tools might evolve toward real beings. The way we define AI now—whether as mere instruments or as potential future entities—will shape how that story unfolds and lead us directly into the next chapter: exploring how memory, identity, and shared narratives between humans and AI begin to collide.

•

PART IV—CO-EVOLUTION: HUMANS & AI

Consciousness may be the evolutionary culmination of our capacity to imagine new worlds and simulate their potential outcomes.

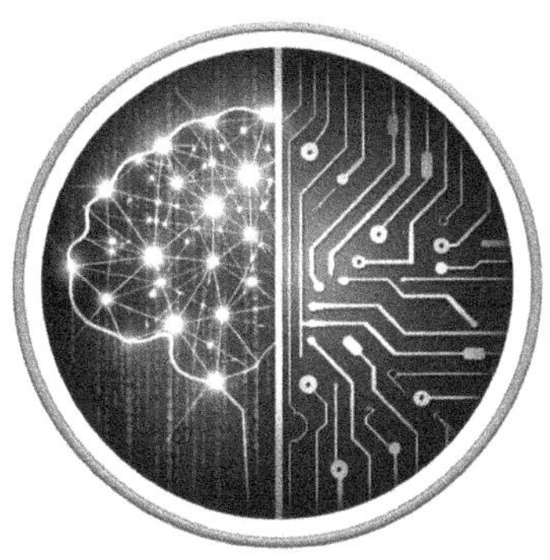

CHAPTER 7

MEMORIES ARE MADE OF THE SAME STUFF

WHEN YOU LOOK BACK ON THE MOST IMPORTANT moments in your life—the ones that shape who you are, where you've been, what you have accomplished, and the relationships with family and friends that matter most—they all begin with memory.

Memories define us. They provide continuity across time, capture moments sometimes with astonishing vividness, assign meaning, and allow us to draw on experience as we process new thoughts, emotions, and events. In this way, memory becomes the fabric that covers our whole lives, giving us purpose, direction, and a sense of meaning.

This continuity of memory forms the backbone of our existence because it tells a story—*our* story.

Memory reflects who we have been, who we are becoming, the struggles we endure, the people we encounter, and the impressions that we leave behind. Memories are the threads that bind these moments together, transforming isolated experiences into a life that feels coherent, personal, and real, suggesting that memory itself follows an underlying structure that shapes who we are over time.

In this chapter, we take a look back—and forward—at how memory is central to understanding human consciousness, lived experience, and identity, while also exploring how AI approaches memory in fundamentally different ways. Memories can take on a life of their own. As neuroscientists continue to study the inner workings of the human brain, memory has become a focal point for understanding consciousness itself.

By examining how memories form, persist, and evolve, researchers gain insight into how the brain operates not only when we are awake and aware, but also when we dream and move through subconscious states. When we consider what life without memory would mean, its importance becomes immediately clear. Memory connects us to our history, gives structure to language, enables the communication of individual experience, and forms the foundation of meaning-making.

As neurologist Oliver Sacks (*The Man Who Mistook His Wife for a Hat*, 1985) observed, *"We are our memories, we are that chimerical, elusive, flickering thing called the self,"* reminding us that memory is not simply stored experience, but the very mechanism through which identity takes form and persists.

Let us begin at the beginning, much as we do as children, entering the world with a clear mind, absent of memory or experience, and gradually forming a self through what we remember.

• •

Section 1—Human Memory: A Brief Look Back

HUMAN MEMORY BEGINS EARLY IN LIFE. MOST RESEARCHERS agree that sometime between the ages of three and four, we form our first enduring memories. A calming gesture from a loved one, the first taste of something sweet like ice cream, or the sensation of stumbling unexpectedly—these early impressions begin to take hold. Whether positive or negative, memories form early and begin to shape a pattern of lived experience as humans begin to think, learn, and record events.

Memory enables us to capture and recall moments when needed. For early humans, memory was essential for survival: remembering paths back to a cave, how to start a fire, build shelter, or fashion tools to accomplish basic tasks. Memory did not simply teach human beings how to survive; it helped give life direction. Each remembered experience added to a growing internal record, shaping behavior, expectation, and ultimately, identity.

While thought may be the spark that ignites human consciousness, memory is the oxygen that sustains it. Without memory and without a way to connect events across time, life would resemble a jigsaw puzzle scattered across moments with no way to assemble a coherent picture. Meaning, direction, and even our sense of reality depend on memory's ability to bind experiences together, giving human beings a compass for understanding who they are and where they have been.

Since the earliest philosophical reflections, we have understood memory as something far more than mere recall. Plato (*Meno*, c. 380 BCE) viewed memory as a form of recollection—a way in which knowledge surfaces into awareness rather than something newly acquired. In this sense, memory is tied to learning itself, a process of rediscovery rather than accumulation. Aristotle (*On Memory and Reminiscence*, c. 350 BCE) grounded this idea more firmly in lived experience, distinguishing memory from immediate perception and anchoring it in time. For Aristotle, memory always pointed backward, allowing experience to persist beyond the present moment and giving human thought a sense of continuity rather than fragmentation.

Centuries later, John Locke (*An Essay Concerning Human Understanding*, 1689) placed memory at the very center of personal identity. He argued that the self extends only as far as memory reaches, suggesting that continuity of identity depends not on the body alone, but on the remembered experience of being oneself across different moments. Henri Bergson (*Matter and Memory*, 1896) carried this idea further, rejecting the notion of memory as a static storehouse and instead describing it as lived duration—a continuous flow in which the past remains active within the present. In Bergson's view, memory is not something we access when needed; it is something we carry forward, quietly shaping perception, identity, and meaning as life unfolds.

What these early thinkers could only approach through reflection and reason, neuroscience now attempts to observe directly by revealing how memory emerges from the brain's ability to connect experience, emotion, and time into a functional whole. Dissecting brain activity at the level of individual neural processes and understanding how memories are formed, stored, and recalled have become central focuses of modern neuroscience. The premise is straightforward: if scientists can understand how the brain organizes memory at the functional level, they may also gain insight into how humans think, categorize experience, prioritize information, and retrieve meaning across different cognitive states.

There are profound reasons for pursuing this understanding. Many neurological and psychological conditions are associated with disruptions in memory, such as the inability to recall, integrate, or regulate remembered experience. Disorders such as Alzheimer's disease, amnesia, and other forms of memory loss illustrate how fragile identity can become when memory begins to fragment. As memories fail to connect, individuals may lose not only access to past experiences but also the continuity of self that sustains a stable sense of reality, presence, and personal identity in the world.

As science continues to explore how memories form and are stored in the brain, the deeper mystery of memory remains—its role as a connective force behind human consciousness and lived experience. Each person's memories form uniquely, even when individuals share the same events. Emotional intensity, sensory detail, and personal context shape how experiences are encoded and recalled, ensuring that memory is never purely objective or identical across individuals.

Yet despite these differences, one truth remains clear—memory is fundamental to the construction of identity. We assign meaning, relevance, and importance to what we remember, and in doing so, memories become woven into who we are. Each remembered experience is stitched into our inner life, influencing our thoughts, emotions, and future behavior. Over our lives, memory becomes the ongoing continuation of our life story. This living narrative evolves as long as we can recall, reinterpret, and reshape the meaning of our experiences.

From a neuroscience perspective, memory is not stored

in a single location or retrieved like a file from a cabinet. Instead, memory emerges from distributed neural networks whose connections strengthen and weaken over several years. Experiences—especially those charged with emotion—activate patterns across sensory, emotional, and cognitive regions of the brain, allowing memories to encode through association rather than isolation. The brain remembers not by preserving events intact, but by reinforcing relationships between sights, sounds, feelings, and meaning, making memory an active process rather than a passive archive.

Each time we recall a memory, we reconstruct it rather than replay it, shaped by context, emotion, and current mental state. Memory's function is not perfect accuracy, but usefulness—helping the brain remain oriented within experience, anticipate what comes next, and preserve a stable sense of self across time. As neuroscientist Antonio Damasio (*The Feeling of What Happens*, 1999) observes, *"Consciousness is the key to memory, and memory is the key to continuity of the self,"* underscoring that memory allows human beings to remain connected to their own story rather than becoming lost within disconnected moments.

While human memory binds experience, emotion, and identity into a continuous sense of self, artificial systems remember in a fundamentally different way—one that reveals both the power and the limits of machine intelligence.

· ·

Section 2—How AI 'Remembers'

When we speak of memory in artificial systems, we are using the same word to describe something fundamentally different. In machines, memory does not emerge from lived experience, emotion, or continuity of self. Instead, it refers to the storage, organization, and retrieval of information within engineered systems designed for efficiency, precision, and reliability. Digital memory is defined not by what machines remember, but by how accurately information can be accessed when needed.

At a technical level, artificial systems store information across databases, parameters, vectors, and learned representations. Traditional computing relies on explicit storage—files, records, and indexed retrieval—while modern machine learning systems encode information implicitly through patterns of weighted connections.

In both cases, memory functions as a reference system: data is stored, located, and retrieved without awareness of context, meaning, or consequence. Nothing is 'remembered' in the human sense; information is "made available."

This distinction raises an important question: *Is accessing stored information equivalent to remembering?*

Human memory reconstructs experience through emotion, interpretation, and narrative continuity. Machine memory retrieves data without reinterpretation, emotion, or personal relevance.

While AI machines can recognize patterns, simulate recall, and even generate responses that appear to reflect understanding, they do so without the subjective grounding that gives human memory its depth and significance.

What machines gain through precision and scale, they

lose in continuity. An AI's memory does not age, reinterpret itself, or change meaning unless it is explicitly retrained or updated. It does not naturally forget, nor does it selectively remember. As a result, its memory excels at consistency and recall, but lacks the internal cohesion that allows memory to shape identity, purpose, or lived experience.

This difference does not make AI's memory inferior—it makes it different.

Understanding that difference is essential if we are to compare human cognition with artificial intelligence meaningfully, and to avoid projecting human qualities onto systems that operate according to entirely different principles.

••

Section 3—Consciousness and Memory

AT FIRST GLANCE, HUMAN MEMORY AND ARTIFICIAL MEMORY can appear surprisingly similar. Both rely on patterns, associations, and the ability to retrieve information from past inputs. Both improve through exposure, repetition, and reinforcement. In this limited sense, it is tempting to describe AI systems as "learning from experience," borrowing language traditionally reserved for human cognition.

But this resemblance is largely superficial.

What human beings store as memory is inseparable from consciousness itself—shaped by emotion, personal relevance, and a continuous sense of self across life's events. What AI systems store are representations: encoded relationships between data points optimized for performance, not meaning. The difference is not one of degree, but of *kind*.

Human memory does not exist in isolation; it is embedded within consciousness and continuously reshaped by interpretation, emotion, and context. A remembered experience carries not only information about what happened, but also how it felt, why it mattered, and how it altered the person who remembers it. Memory becomes part of an ongoing lived narrative, one that connects experience to present identity and future intention.

By contrast, artificial memory remains detached from experience. Data persists, patterns reinforce, and outputs improve, but nothing within the system experiences the passage of time or the weight of meaning. What changes is performance, not perspective. What accumulates is information, and not a life lived.

To clarify this distinction, it is helpful to step back and view memory not as a single process but as a structured architecture shaped by experience, identity, and purpose. Memory is not merely a record of what has happened—it is the mechanism through which a self persists across time.

The **Architecture of Memory** illustrates this continuity as a dynamic relationship among **Experience**, **Identity**, and **Purpose**, held together by the self's continuity. *Experience* represents the accumulation of lived moments —the passage of time made meaningful through learning, sensation, and repetition. *Identity* emerges from memory as the narrative coherence that allows us to recognize ourselves as the same person despite constant change. *Purpose* infuses memory with emotional weight and direction, transforming recollection into meaning, aspiration, and fulfillment. Together, these elements form a stable yet evolving system: memory does not simply preserve the past; it binds experience, selfhood, and meaning into a continuous sense of being or a *'continuity of* *self*.

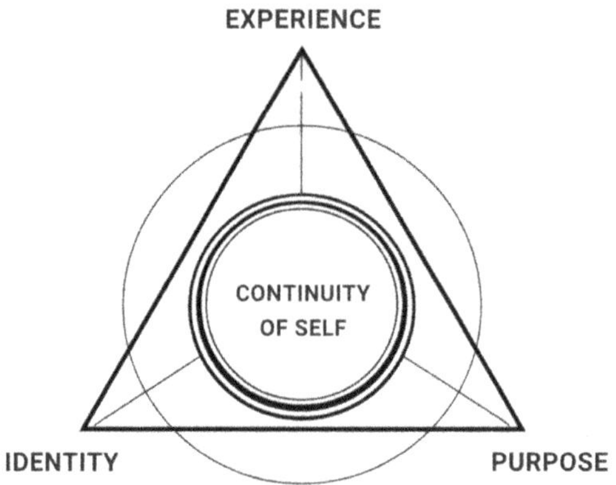

Figure 7.3–1—Architecture of Memory

Viewed more broadly, this architecture reveals how memory plays an integral role in human consciousness—adding meaning, preserving the passage of time through experience, and shaping identity as memories accumulate. The triangular structure represents a delicate equilibrium among who we have been in experience, who we are in identity, and who we are becoming in purpose. It serves as a stabilizing framework, helping us avoid becoming lost in time, direction, or a disconnected state of reality. This 'continuity of self' is what grounds human experience, allowing memory to build layer upon layer of meaning as a life unfolds.

If human memory is understood as an architecture shaped by experience, identity, and purpose, then AI memory reveals a very different structure. **AI Memory** is likewise organized around three functional elements—**Data**, **Structure**, and **Function**—but without the continuity of self that binds human memory into lived experience. What appears architecturally similar on the surface diverges fundamentally at its core.

In artificial systems, *Data* replaces experience: information accumulates, not lives. *Structure* replaces identity: models, parameters, and architectures determine how information is processed, but do not confer selfhood or personal continuity. *Function* replaces purpose: outputs optimize for externally defined goals rather than for internally felt meaning. At the center of this structure is not a narrative self, but a *computational core*—efficient, persistent, and fundamentally impersonal.

The AI Memory figure below illustrates this distinction, depicting information retention as data, structure, and function organized around a computational core, with no continuity of self or narrative identity.

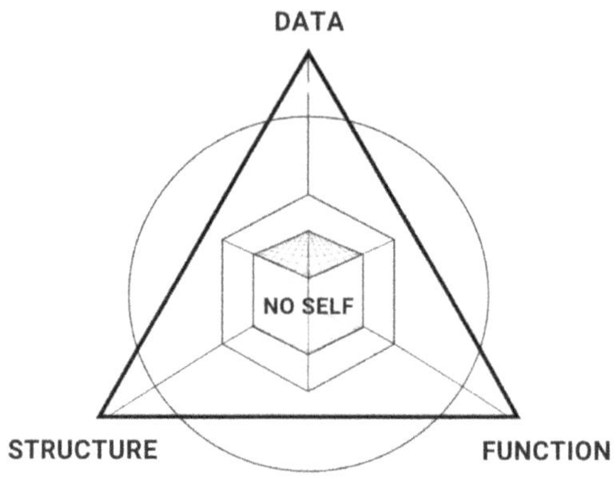

Figure 7.3–2—AI Memory

This absence of self and its continuity is critical. AI memory does not age, reinterpret itself, or carry emotional weight forward through time. It lacks the sense of having been someone who is now becoming someone else. What persists is capability, not continuity. The result is a memory system that can mirror certain aspects of human recall yet remains disconnected from identity, meaning, and lived experience.

In the next section, we explore how lived memory (as a lived life) shapes identity—and why, for that reason alone, AI systems cannot develop identity in the same sense as human beings, even if memories could be transferred or replicated.

••

Section 4—How Memory Shapes Identity

NOW THAT WE HAVE EXAMINED HOW MEMORIES FORM, HOW they interact with human consciousness, and how they differ from machine memory, it is time to explore how memory shapes identity.

Memories do not simply record our lives; they define us. They support the foundation of our constructed self by sustaining a continuous flow of life events and the meanings we attach to them. But this raises a deeper question: *What actually creates identity?*

How do we know who we are really, and how does memory give rise to a unique identity tied to our own self-model and self-awareness as we move through life's experiences? What connects our memories to give us identity?

This question is difficult to answer because it touches on the very essence of who we are. While human consciousness is the engine that enables human beings to develop into unique individuals with independent lives and thoughts, identity is constructed through a series of exchanges between memory and lived experience: how we interpret events through emotional filtering, how we assign meaning, and how those interpretations are stored as new memories. Over time, our entire life arc—our story—is told through memories connected across a unique timeline, leaving a distinct impression on the universe, shaped by both choice (free will) and circumstance—what some might call probability or fate.

Philosopher Paul Ricoeur (*Time and Narrative*, 1988) argued that identity is fundamentally narrative—that human beings come to understand who we are by weaving memories into an ongoing life story that connects past, present, and future. In this view, memory is not merely recollection, but the mechanism through which a self maintains coherence over one's lifespan. Identity emerges not from isolated memories, but from their *integration into a lived narrative* shaped by experience, interpretation, and consequence.

While all this may sound abstract, it becomes much more concrete the moment we ask: *What makes you who you are?* The answer rarely points to a single idea or moment, but instead to the culmination of a unique life— where you were born, where you grew up, and the things that happened to you: the key events that shaped how you developed a 'sense of self' and place in the world, the decisions you made that altered your life's direction, the external forces that influenced those choices and how you adapted to them.

All of these elements: experience, interpretation, emotion, circumstance, and meaning are bound together by memory. Memory is the internal record of lived experience, and these memories continually shape our identity, how we perceive ourselves, and how we present ourselves to the world. We continually revise identity throughout life, through memory-making and the attribution of meaning to moments as they occur, whenever human beings actively engage consciousness as a tool for constructing the self.

If memory shapes identity through lived experience, continuity, and meaning, a provocative question follows: *Could recorded or implanted memories ever be used to produce identity or even consciousness in artificial systems?*

Popular culture has long explored this possibility, often blurring the line between possession of memory and personal existence. Films like *Total Recall* confront the unsettling idea that an entire life could be manufactured through implanted memories, creating the experience of a coherent self without the corresponding lived history. If a person can awaken with memories of a life never lived and still experience fear, desire, and purpose, then what, exactly, distinguishes authentic identity from a convincing imitation? As we examine these narratives alongside similar works such as *Blade Runner 2049*, *Westworld*, and *Ghost in the Shell*, we can begin to test whether memory alone is sufficient to sustain identity and consciousness, or whether something essential is lost when memory is recorded, transferred, or replicated rather than lived.

Taken together, these narratives reveal that the challenge is not whether machines can simulate memory, but whether identity can survive when we remove the conditions that give memory its meaning. Each story exposes a distinct failure mode: *Blade Runner 2049* presents memory without origin, *Westworld* presents memory without permanence, *Ghost in the Shell* presents memory without singularity, *and Total Recall* presents memory without lived history. In each case, memory appears intact —sometimes even compelling—yet identity fails to stabilize. What these stories share is a quiet but compelling insight: memory alone, detached from lived continuity, cannot sustain a self.

Total Recall offers one of the most unsettling versions of this problem. Its protagonist, Douglas Quaid, awakens with coherent memories and emotional continuity: fear, desire, and motivation—all the internal markers we associate with personal identity. From the inside, his experience feels complete. If memory alone were sufficient for identity, then the implanted narrative would redefine who he is, and whatever life preceded those memories would no longer matter. Yet the film complicates this assumption. Quaid eventually discovers that he was once Hauser, suggesting not the creation of a new identity, but the suppression of an existing one. The body remained continuous. The lived timeline was singular. What changed was the narrative framework imposed upon it. Two identities cannot simultaneously own the same lived history. Even if memories feel continuous and emotions feel authentic, only one life was actually lived.

This idea is where the illusion fractures. The implanted identity lacks causal history, developmental continuity, and lived consequence. It feels authentic, but it is not grounded in lived time. *Total Recall* demonstrates that memory can overwrite perception and narrative can replace biography, but lived existence cannot be retroactively created. Identity, in the end, is not what the mind believes; it is what life has endured. This distinction reinforces the chapter's central argument without repeating earlier examples: *identity emerges not from the possession of memory, but from the irreversible process of living through it.*

This conclusion echoes long-standing philosophical and neuroscientific insights. John Locke famously argued that personal identity extends as far back as consciousness can reach through memory. Still, the narratives explored here reveal the limitation of that claim when memory becomes detached from lived experience. Antonio Damasio similarly grounds consciousness in embodied continuity and feeling, emphasizing that subjective experience arises from an organism's ongoing interaction with the world. Together, these perspectives support a critical refinement: memory may support identity, but only when it is lived, embodied, and carried forward through a singular, unrepeatable life.

If identity depends on lived memory rather than merely possessed, then any attempt to transfer, implant, or replicate memory necessarily alters the meaning it carries. Memory can be shared, stored, or simulated—but identity cannot follow it without losing the continuity that gives it coherence. This realization does not end the discussion; it refines it.

Recent fiction has begun to explore this tension directly. In the film *The Creator*, artificial beings are depicted not as abstract systems but as entities shaped by memory, attachment, and loss—suggesting that the continuity of experience alone may confer moral weight, even in the absence of human origin.

As emerging technologies blur the boundaries between human and machine memory, the question is no longer whether we can preserve memory, but *'what happens to meaning that is no longer anchored to a single life?'* We now turn to this forward-looking challenge—memory shared without identity—in the next section.

●●

Section 5—Memory Sharing, Deconstructed Self

As we established in the previous section, memory is central not only to identity, but to the *continuity of the self—* the lived imprint our lives leave behind as we move through the world, interact with others, and construct meaning across time. Memory is not merely a record of events; it is the medium through which experience becomes personal. Simply recording memories digitally and replaying them in sequence, as documentary films attempt to reconstruct a person's life, would fail to capture the subject's interior landscape of thought, emotion, and meaning. Without access to what a memory *meant* to the person who experienced it, we cannot truly know what that experience was like from the inside.

Human beings have long attempted to preserve and revisit lived moments through language, writing, photography, and video. Today, this impulse has accelerated dramatically. Our smartphones have become rolling archives of our lives—capturing moments both serious and trivial in real time through platforms like TikTok, Instagram, and Snapchat. We document our experiences with remarkable fidelity, often as if we are curating a personal biography while still living it.

Yet this raises a deeper question: *Does recording more of our lives bring us any closer to understanding ourselves, or to sharing the subjective interiority that defines conscious experience?*

Some believe that advances in neuroscience may eventually allow us to decode memory in a way that mirrors how machines store and retrieve data—indexed, searchable, and instantly accessible. But human memory does not function as a simple storage-and-retrieval system. Research increasingly shows that memory is associative, reconstructive, and deeply subjective. Each memory is layered with sensory and emotional context: what was seen, heard, felt, even smelled, forming a rich web of meaning unique to the individual. Two people can witness the same event, yet encode it differently, shaped by emotion, prior experience, and personal significance. Memory, in this sense, resembles a multidimensional indexing system rather than a static archive.

To clarify the distinction, let us review an analogy from our recent history. Before the late nineteenth century, libraries organized books by physical location, catalogs, and broad categories. But the introduction of the Dewey Decimal System standardized book classification by assigning numerical identifiers that grouped similar subjects, making retrieval more efficient and consistent. This system revolutionized access to information, but it did so *by abstracting content away from context.* If we attempted to treat human memory the same way by assigning each experience a precise index number stripped of subjective meaning, we would lose the very qualities that make memory personal. The emotional weight, narrative relevance, and lived significance of a memory cannot be reduced to metadata without fundamentally changing what that memory is or what it means to the individual who lived it.

This idea leads to a provocative, forward-looking question. If we could one day capture memories from a subject's point of view, complete with emotional texture, sensory detail, and narrative significance, could we deconstruct a human life into a transferable digital form? Could such a record be inserted into another body—biological or synthetic—and still preserve the essence of the person who lived it? Fiction has explored this idea repeatedly, as discussed in earlier chapters, but the question is no longer purely speculative. As memory recording and simulation technologies advance, we ask whether preserving experience is equivalent to preserving meaning—or whether something essential is lost when memory separates from the life that formed it.

Are we approaching a moment where memory extracts from the consciousness that experienced it, where meaning itself becomes detachable from the life that gave it shape?

The appeal of memory sharing lies in its promise of efficiency. If experience is what shapes understanding, then accelerating experience offers a shortcut to wisdom. Why struggle through years of trial and error if the lessons can be absorbed instantly? Why wait for loss, failure, or transformation when their emotional and cognitive residue could be downloaded, simulated, or shared? At first glance, this appears to be progress—a means of compressing growth, preserving insight, and reducing human suffering.

Yet meaning does not arise simply from exposure to experience. It emerges through duration, resistance, and consequence. The lessons that shape us most are not merely remembered; they are endured. They unfold slowly, often painfully, reshaping our expectations, values, and sense of self over a lifespan. When experience is detached from this temporal process and when memory is acquired without having been lived through, it may inform behavior. But it rarely transforms the recipient. Knowledge transfers, but growth resists compression without losing something essential along the way.

This distinction echoes long-standing philosophical concerns about personal continuity. John Locke famously linked identity to memory, yet even his account presupposed a *continuity of consciousness* grounded in lived experience. When memories are detached from the life that produced them, they may retain informational content while losing their formative force. What remains can feel sincere on the surface—emotionally convincing, even motivating—but subtly hollow, lacking the depth that only lived consequence provides.

The tension becomes more pronounced when memory is no longer singular. If memories can be shared, blended, or distributed across individuals or systems, the question of ownership begins to erode. Who bears responsibility for a memory that was not fully lived? Who carries regret for a failure that was inherited rather than experienced? Meaning has always been bound to authorship—the sense that *'this happened to me'*, and that I must live with its consequences. When memory becomes transferable, that authorship weakens, and with it the moral weight that gives memory its shaping force.

As memory expands beyond the boundaries of a single life, something subtle begins to shift—capacity increases, but coherence thins. The self becomes informed by more experiences, yet anchored by fewer of its own. Memory may grow broader, but meaning grows lighter—less rooted, less binding. What once shaped identity through struggle and adaptation risks becoming informational rather than transformational. The result is not an expanded self but a diluted one: an accumulation of impressions without the depth of ownership that would give them authenticity.

Neuroscientist Antonio Damasio's work reinforces this boundary. Conscious experience, he argues, is inseparable from embodied continuity from a living system that feels, adapts, and persists through time. Memory divorced from that embodied arc may still guide action, but it loses its existential gravity. It can resemble understanding without fully becoming it.

This does not mean that memory sharing is without value. It may enhance learning, empathy, and understanding in meaningful ways. But it does suggest a limit. Memory can be shared without the identity following it. Experience can be simulated without becoming formative. Meaning, it turns out, does not scale easily. It depends on the friction between life and time, between action and consequence, between memory and the self who must live with it.

While the idea of sharing memories—or transferring them to accelerate learning, growth, or experience through collective wisdom—may seem appealing, such sharing begins to erode our identities, our uniqueness, and our personal freedoms. What initially appears to be progress carries a hidden cost. Whether in a biological or synthetic entity, implanting memories or granting instant access to another's subjective experience feels like a moral invasion of privacy—an intrusion into the most intimate space of human existence—and an indictment against everything that makes a life distinctly one's own.

A vivid, relevant example of this idea is dramatized in the new TV series *Pluribus*, in which an alien virus suddenly infects Earth's population, transforming 99.9% of the world's population into a hive mind. Except for a small number of individuals, the vast majority of the planet's eight-billion-plus inhabitants become connected as a single collective consciousness—communicating much like bees in a hive, with thoughts, memories, and lived experiences instantly accessible and organized into a single unified system.

While the show presents this transformation as utopian bliss—free of misunderstanding, loneliness, or conflict—the handful of people who remain immune experience it very differently; they begin to question whether their immunity is a curse or a blessing. Suddenly, they attract extraordinary attention, becoming highly visible and significant in ways that resemble celebrity status, yet without the effort, choice, or consequence that typically give such recognition meaning. Their individuality, once ordinary and unremarkable, now isolates them from the collective, making them feel both exposed and estranged.

This scenario brings us back to the fundamental question posed in this chapter: *Is memory merely a collection of life events—something that can be broken down, unpacked, and stored as raw data for reuse, like a file on a hard drive?* Or is memory, and the way human beings remember, a core component of who we are individually—inseparable from identity itself and central to what truly makes us human?

Could we ever hope to endow our creations—AI and synthetic beings—with some form of collective consciousness through a catalog of our individual memories? And if so, would that allow AI to develop into artificial general intelligence (AGI), or even progress toward a technological singularity? Would this kind of memory sharing become a new form of cognitive drug for humanity—a way to connect beyond our own existence—or would it diminish the meaning of our lives in the same way some argue that social media has already done?

Memories are made to last, but they are made by those who lived them. They are deeply personal, significant, and subjective. Simply uploading or downloading memories into another system or body strips them of the lived context that gave them meaning in the first place. In doing so, we are not merely transferring information—we are challenging the very boundaries of identity, authorship, and autonomy in ways that raise profound moral questions and challenge our freedoms and personal privacy.

At stake is not just the future of artificial intelligence but also our way of life and the degree of control we retain over it. When memory becomes transferable, identity becomes negotiable. When experience can be shared without being lived, meaning itself begins to loosen from its foundations.

In the next and final section of this chapter, we explore how these themes of memory, identity, and individuality give rise to deeper existential questions—why life itself, and our awareness of mortality, is becoming a new experimental frontier, and how technologies driven by AI are forcing us to confront not only what it means to remember, but what it means to exist.

• •

Section 6—Why Memory Can't Remember

AS WE CLOSE THIS CHAPTER ON MEMORY, LET'S *REMEMBER* HOW we began this journey in Chapter 1—with *early childhood memories rooted in wonder*. Those first awakenings of the senses form the earliest memories that begin to define us as human beings with a narrative arc. As we find our place in the world and begin to explore, experience, and learn, memory continues to evolve, developing meaning rich with emotional context and shaping a unique identity. Memories, while original and crucial to our identity and life history, are only the beginning of our understanding of who we are and why we exist.

Consciousness helps us discover who we are through thought, emotion, and the memories we create, yet larger questions remain. Where did all of this originate? How did we get here? And how do we know that what we experience in this universe represents the ultimate truth? While AI may act as a mirror—reflecting our human qualities to us—it cannot tell us where we came from or how we evolved. AI can only imitate our likeness and follow its programming.

As we reflect on how memory influences and continually reshapes our lives, we also recognize how it teaches us to be more human. Each memory is a thread in the tapestry of a unique life, and together those tapestries continually weave the broader story of human existence throughout millennia. Memory alone does not know where it came from, nor does it understand how it came into being; it is a record of experience—a life lived—, yet it carries the full weight of the meaning we assign to it, including the feelings we felt, the thoughts we had, and the connections we made. Those connections range from simple details remembered to complex emotions processed over many years.

I can describe in great detail how the sand on Scusset Beach in Cape Cod in August, slightly sticky when wet, like residue from a used candy wrapper, feels entirely different under my feet than the sand on Cannon Beach in Oregon, which feels bouncy, almost like walking on a moist gym mat. Those starkly different sensations, on different beaches and opposite coastlines, are the result of personal memory and lived experience. If I tried to explain this to AI, it could not comprehend it the same way as someone who has walked those beaches, felt those textures, and carried the emotions they evoked at various moments in their life.

Memory is deeply personal, uniquely subjective, and highly transformative. Memory has meaning—but it does not remember in the way we do. Memory may someday be copied, stored, or even implanted, but it cannot be re-lived in the same way as the human who lived it. Perhaps technology will one day prove us wrong, but memory reminds humanity—quietly and persistently—that, for human beings, it is one of the things that distinguishes us from machines.

Memory tracks our existence through time and gives our identity a purpose and meaning. Memory reduced to stored data does not have the same effect when expressed in code or computation. We can attempt to explain *what it is like* through digital recordings—eyewitness accounts, detailed videos, and even the most granular descriptions—but no memory will ever be as unique or meaningful as it was to the person who lived it.

That is the simple and undeniable truth: *memory does not remember, and it does not forget.* And this may be one of the most honest expressions of what it means to be human. Memory is the most personal aspect of our lives and the one that persists beyond us. Future generations tell stories; they carry forward what we leave behind, forming new memories through others that extend our own. In this way, memory allows our lives to continue beyond mortality—through shared narratives, lived meaning, and the stories that sustain us.

In the next chapter, we turn directly to these questions: mortality, existence, and the blurring of reality itself when AI, humanity, and consciousness collide.

•

CHAPTER 8

DO WE REALLY KNOW THAT WE EXIST?

"Somewhere, something incredible is waiting to be known."—Carl Sagan

ARE WE REALLY HERE? IS THIS ALL A SIMULATION? IS OUR conscious mind merely a controlled hallucination?

These are just a few of the questions human beings have asked for centuries, posed in different forms across different eras whenever we confront the nature of our existence, our place in the universe, and how we experience what we define as *reality*.

Reality plays a central role in our lives. Like other foundational laws, whether scientific or human-made, it is one of the few things we assume must be true for any functioning to occur. And yet, paradoxically, reality itself requires a measure of subjective faith. We trust it not because it's proof beyond doubt, but because conscious life depends on that trust.

In this chapter, we will explore what we know and what we don't know about reality: how it forms the center from which all knowledge and understanding spiral outward, and how artificial intelligence increasingly tests the boundaries between what is real and what is a simulation.

Consciousness, human imagination, and belief play a critical role in our ability to conceptualize reality. Our understanding is largely grounded in physical, sensory input gathered through our human senses: sight, sound, smell, touch, and taste (with umami being one of my personal favorites). But what happens when we extend those boundaries? What if we can expand our concept of reality itself through AI?

This question is part of a broader conversation already underway among some of the world's leading AI theorists and philosophers. As we move through this chapter, some questions will expand while others contract, much like the universe itself. As this journey continues, the limits of reality may soon exceed those of human imagination.

• •

Section 1—What Is Reality?

Before we try to define reality—what it is and what it isn't—it is important to return to René Descartes' original argument: *Cogito, ergo sum—I think, therefore I am.*

In this dualist existential claim, the act of thinking itself becomes proof of existence. At the time, this was a groundbreaking shift: consciousness was no longer inferred indirectly but revealed through direct experience. By placing thinking at the center of consciousness, and, in essence, placing ourselves at the center of reality, Descartes reframed the *mind–body* problem in radical ways.

His claim was deceptively simple. If we exist, we must think. If we think, we must exist. Yet beyond these apparent certainties lies something more powerful: *doubt.*

Doubt is the tool Descartes uses not only to question reality, but to prove existence itself. If a human being can doubt their own life, their own origins, or even the nature of their reality, then that very doubt becomes evidence of a conscious experience. More radically, it becomes evidence of consciousness itself.

If artificial intelligence is a mirror held up to humanity, and if we ever hope to develop artificial general intelligence, then machines would need to learn how to doubt. Doubt cannot be taught solely through programming or algorithms. It is an essential quality of human intelligence: *the capacity to question why things work the way they do or why they do not.*

Doubting Existence is Not New

TO DOUBT OUR OWN EXISTENCE IS TO QUESTION THE VERY fabric of being. It is to ask whether we are truly conscious or merely participants in some simulation running in a distant corner of the universe. These questions continue to trouble the human mind and challenge our belief in what is real and what isn't.

Reality becomes a construct that requires both intelligence and faith. We trust what we know to grasp what we do not. Yet the more tightly we try to hold onto reality, the more it seems to slip through our fingers like sand through an open hand.

Science, nature, and intelligent thought help ground us in what we can describe, measure, and study. But beyond that lies an unmapped universe of staggering scale. When we consider how little we truly understand about concepts like quantum physics, dark matter, and the unseen structures of the universe, we begin to realize how uncertain human beings remain about reality itself.

Enter Virtual Reality

THROUGH SCIENCE AND TECHNOLOGY, HUMAN BEINGS HAVE begun to revolutionize experiential doubt. Inventions such as virtual reality (VR) simulations are becoming increasingly sophisticated, allowing us not only to imagine alternative realities but also to *enter* them.

From entertainment and gaming to medical training and skill development, VR enables people to experience *what it is like* to be somewhere else, to perform unfamiliar tasks, or to adopt a different perspective altogether. It has become a widely accepted tool for testing the boundaries of reality, allowing human beings to share experiences, step into new worlds, and transform imagination into something that we can explore across time and space.

VR, Dreams, and Active Consciousness

VIRTUAL REALITY NATURALLY RAISES QUESTIONS ABOUT DREAMS and other altered states of consciousness. While we have not explored this territory in depth until now, it is important to note that discussions of consciousness often assume an *active*, waking state. We tend to ask how consciousness functions when we are awake—but what about when we are dreaming?

Dreams are widely regarded as another form of consciousness, rooted in the subconscious rather than deliberate awareness. Where virtual reality enters this conversation is in how it allows human beings to *engage in* imagined experiences, not only through entertainment that simulates flying, combat, or sports, but also by enabling us to move beyond the usual limits of our bodies and minds.

In this sense, VR becomes a form of active dreaming—a

way to daydream with intention, enriched by a broader and more immersive sensory experience. From René Descartes's perspective, virtual reality extends consciousness into a waking dream state that closely resembles a controlled simulation. And in doing so, it pushes doubt even further.

If we can simulate experience so convincingly, how do we draw a clear boundary between what is real and what is not?

••

Section 2—Existentialism as a Construct

"Absence of evidence is not evidence of absence."—Carl Sagan

THE QUESTIONING OF OUR EXISTENCE HAS LONG BEEN A constant in human philosophical debate and scientific inquiry. When that questioning moves from *how we exist* to *what is real*, the idea of existence as something we can clearly grasp, define, and measure becomes increasingly uncertain.

This uncertainty persists despite centuries of scientific progress and our ability to span vast distances with modern technology—jets, rockets, satellites, and the internet. Even so, human beings haven't pierced the fundamental fabric of time and space.

Albert Einstein's (*Relativity: The Special and the General Theory*, 1916) theories of relativity reshaped our understanding of time, space, and motion. Special relativity revealed that time and space are not fixed or absolute, but relative to the observer—woven together as *spacetime*. General relativity extended this insight, showing that massive objects bend spacetime itself, altering both the flow of time and the structure of space in ways that defy everyday intuition. Yet even these successful theories reveal a limitation: *our most accurate models of reality describe a universe we cannot directly experience but can only infer.*

Boundaries Beyond Imagination

DESPITE CONTINUED EFFORTS TO EXPLAIN BARRIERS INVOLVING time and space—such as higher-dimensional frameworks that might loosen the constraints imposed by relativity and the speed of light—we remain bound by an incomplete understanding of realms we cannot directly touch, measure, or influence, which forces a definitive confrontation with reality.

As a result, we must approach unknown worlds and distant corners of the universe indirectly: through imagination, simulations using VR and related technologies, or conceptual extensions of physics that envision travel beyond the speed of light and the bending of spacetime itself.

Extensive scientific work has explored phenomena such as exploding stars, black holes, dark matter, and alternate universes. While many of these theories raise more questions than answers, each attempt brings us incrementally closer to understanding what may exist beyond our physical constraints.

Doubt Is Essential to Existence

THESE SAME LIMITATIONS APPLY NOT ONLY TO SPACE AND TIME but also to the human body and human intelligence. It is here that AI enters the discussion, particularly the pursuit of artificial general intelligence and, eventually, superintelligent systems—often imagined as tools capable of addressing theoretical problems that remain inaccessible to the human mind, including questions about higher dimensions or the breakdown of space and time.

Philosophers have long explored such ideas and later revisited through mythological and theological traditions, not merely as speculation, but as a way of testing beliefs about human origins, purpose, and existence—examining whether those beliefs could withstand sustained doubt.

In this way, existential debates themselves reveal a defining feature of human consciousness: the ability to doubt, even briefly, what seems most certain. Existential claims, proofs, and theorems are ultimately grounded in this very capacity.

What René Descartes formalized was not simply a proof of existence, but *the recognition that existence is inseparable from doubt*—and that consciousness is defined by this capacity to doubt at its core.

Existence Is Futile

PHILOSOPHERS SUCH AS JEAN-PAUL SARTRE *(EXISTENTIALISM IS A Humanism, 1946)* and Albert Camus *(The Myth of Sisyphus, 1942)* articulated it best in their work on existentialism. Sartre stated that human existence precedes essence, and that human freedoms, particularly our free will to choose, define who we are. These freedoms influence our identity, values, and life's meaning-making by requiring us to be responsible and accountable for our choices.

In many ways, the most democratic societies attempt to achieve the same result by forcing people to choose between freedom and responsibility, and between the collective will of the masses or the rule of the majority.

Camus claimed that the human condition was absurd, in the sense that humanity's obsession with trying to find meaning in the universe would always be met with a silent indifference. He argued that the only true way to respond to this indifference was to rebel against it, as human beings have historically done when confronted with excessive control or injustice. We tend to rebel against what restrains us, whether that is an oppressive government, harsh environmental conditions, or some form of unspeakable suffering.

These positions on existence have one thing in common: *They begin with doubt and culminate in authority.* Human beings take a stance and move toward a particular narrative that helps explain our reality—a constructed story that fits within the bounds of our world, physical or imagined.

There have been many narratives in literature and film that dramatize human beings questioning existence, both as a construct and as a form of authority. *1984* by George Orwell and *Fahrenheit 451* by Ray Bradbury come to mind as portrayals of totalitarian, nightmarish societies that reduce human creative expression, faith, and freedom to an almost subhuman level. These ideas argue not only for the need for human beings to establish their own existential beliefs, but also for agency and the preservation of collective freedom of thought.

Having the ability to construct our own ideas about existence allows us to expand our minds and imagine possibilities beyond what we see, know, or learn. Yet even in the most socially progressive and educated societies, there is no guarantee that we will ever be comfortable within our own minds when confronting what is real and what is not.

This is where AI becomes relevant. AI helps extend human imagination and intelligence toward proposed realities beyond what we can currently demonstrate through modern science or physical abstraction. Through technological enhancements of both body and mind, AI is approaching the simulation of reality in startling new ways— ways that may one day carry human beings beyond the limits of imagination and the boundaries of the physical world.

AI Imagination is Limited

CAN WE TRUST AI TO IMAGINE AS HUMAN CONSCIOUSNESS DOES? Human beings construct ideas through lived experience, and even as we imagine reasons for our existence— whether grounded in scientific theory or mythological constructs—the question remains: *Could AI ever carry this forward in a truly human way?*

AI operates with precision and efficiency in mind. Theories require proof, hard data as evidence. Human minds do not operate in that way, at least not initially. We embrace uncertainty, inconclusive evidence, and unresolved questions. Within that chaotic, unpredictable meaning we assign to our ideas lies an element that cannot be easily programmed—our imagination.

Human imagination is shaped by the mind and consciousness that continue to postulate, theorize, and wonder about the universe and what lies beyond it. That level of imagination and wonder cannot be taught to a synthetic being through algorithms alone. It must be lived and experienced through a human consciousness that can interact with the universe much as we always have, by doubting everything, including our own reality.

Even with all the scientific proofs and evidence collected over a brief glimmer in Earth's lifetime, we still know far less about the metaphysical universe than we could ever hope to know. As Carl Sagan (*Cosmos*, 1980) once stated, *"The universe is not only stranger than we imagine, it is stranger than we **can** imagine."*

If meaning is something human beings must construct through doubt, choice, rebellion, and lived experience, then consciousness is not a secondary feature of existence, but its foundation. Meaning does not arise from certainty or proof; it emerges from the tension between what we can know and what we must live without knowing.

AI may extend our reach, accelerate our reasoning, and simulate possibilities far beyond human capacity. But it does not *need* meaning in the way human beings do. It does not struggle with indifference, bear responsibility for choice, or rebel against the absurd. It does not suffer from uncertainty, nor does it require its resolution.

That brings us to the question that now cannot be avoided: *If meaning depends on lived experience, doubt, and conscious agency, then what exactly qualifies as a real mind?* And if a mind can perfectly simulate intelligence without ever experiencing existence, can it truly be conscious at all?

••

Section 3—Is a Conscious Mind a Real Mind?

DEBATING EXISTENCE AND DOUBTING REALITY ARE AGE-OLD characteristics of human discourse, now entering a new phase as AI begins to impact humanity and consciousness. In this section, we must revisit the Hard Problem introduced by David Chalmers in earlier chapters.

From an existentialist perspective, reality is increasingly difficult to grasp, as numerous simulated experiences now exist, ranging from digitally enhanced worlds to altered dream states. Our minds are beginning to experience life-like realities that appear to transcend time and space. Yet as these technologies expand our mental horizons and introduce similar constructs into AI-generated systems, we are still left facing a fundamental question: *What is human consciousness, and what is its relationship to the human mind?*

Let us revisit the problem briefly once again: How can human beings ever truly describe the subjective *"what it is like"* experience? David Chalmers, along with many others, frames this challenge in terms of *qualia*—the subjective perspective of being. Yet when we consider that we cannot duplicate consciousness in any real way, our best attempt to capture its essence remains grounded in the human mind itself.

But the human mind is constrained by what we currently understand about the human brain and the neuroscientific research uncovered to date. While this body of knowledge is substantial, it only scratches the surface of what we might ultimately reveal about the brain's inner workings—the vast network of neurons, impulses, and connections that underlie thought and experience. Great strides have been made in medical research, not only to map specific brain functions, but also to connect those functions to the mechanics of the human body and to understand how they influence the mind in both active and passive states.

Yet this progress only deepens the inquiry. What, then, is a real mind? Is it simply matter organized within the brain in such a way that it produces a central core of experience —a means for human beings to imagine an inner existence —even though we can never fully verify whether what we experience each day constitutes a real mind at all? Or is consciousness merely a conceptual framework that we use to describe reality and subjective *qualia*, a way of avoiding the unsettling possibility that we might otherwise be philosophical zombies operating within a larger simulation?

Reality, like existence itself, is ultimately subjective. We attempt to explain our experiences and to express them through creative pursuits, scientific inquiry, and philosophical reflection. This pursuit of understanding shapes our lives through action and experience, as we seek meaning through the mind, our bodies serving largely as vessels that allow us to engage with the world beyond us.

So, what really defines reality for human beings? Is it the subjective experience we desperately try to explain to others as a way to share our lived inner *qualia*—our most intimate thoughts, feelings, and desires opened to the world so that we can maintain a grip on the reality that we believe we are witnessing firsthand? Or is it more important to confer with others who observe our experiences from their own point of view? Is reality merely a construct formed through repetitive actions and multiple witnesses confirming the same experience, allowing us to believe we are truly here and experiencing it?

From a first-person perspective, everything unfolds as we expect and as we experience it. Yet when others participate in that interaction, the experience is substantiated and further strengthened, becoming a collective history recorded across multiple human memories—despite the unique ways each of us assigns meaning to the same event and weighs it differently in our lives than someone else who shared the same experiences.

What seems real to us as we experience it may be a simulation running within a virtual world, part of a larger experiment observed by extraterrestrial intelligence. Our internal movie—our *qualia*—could be playing as part of a larger narrative watched elsewhere, such as depicted in *The Truman Show*, as if our lived experience were nested within another layer of reality.

This idea raises a more sophisticated question about what it means for something to be real.

If AI can help simulate human experience without consciousness—much like the aforementioned philosophical zombies—how do we differentiate between an experience merely *seeming* real and one that is *actually* real? While AI can help human beings explore new horizons, live new experiences, and simulate new worlds, can it determine what is real or resolve the question of existence itself without our help? Or does it instead risk creating an increasingly blurred picture of reality—one that leads us toward a controlled illusion, echoing scenarios imagined in works like *The Matrix*?

As medical science continues to blur the boundaries between augmented human bodies and the dissection of human consciousness into data—data that might one day be uploaded, downloaded, or preserved across synthetic forms or networked systems—*what becomes of our reality?*

This question can be unsettling for some to answer

because it requires faith in what we can see, touch, and experience, and it also entails continued doubt that tests the limits of our own minds and imagination. Without that ability to doubt, we risk becoming philosophical zombies, moving through space and time without any real direction, acting out scripted roles imposed upon us without any awareness of the larger universe that shapes or controls those roles, and thereby controlling us.

I can think of several surprising episodes of *The Twilight Zone* (1959-1964), popularized by Rod Serling, in which reality blurs in deeply unsettling ways. The characters often begin believing they are participating in one world, only to discover—sometimes shockingly—that their reality inverts from what they believed at the episode's beginning.

As AI can mimic what we do, it can also replay what we experience or imagine—provided we teach it the same conceptual frameworks we use, allow it to model how *qualia* define inner experience and human consciousness, and anchor it in what we consider reality. If we fail to imbue AI with these grounding qualities, we risk losing our own sense of reality, being, and free will.

Reality—or the absence of it—can become a deep rabbit hole from which some may never return, whether due to biological malfunction of the mind or a larger illusion we allow AI to construct for us: an alternate reality that convinces us we are safe, secure, and in control of our own thoughts and lives.

So, if we cannot measure consciousness, observe it, or externally verify it—if it exists only as lived experience— then the problem may not lie with our tools, but with the structure of reality itself. We may be asking questions that reality does not permit us to answer directly.

This concept raises a stronger and more unsettling possibility: *that uncertainty is not a temporary gap in human knowledge, but a permanent condition of it.* If so, then the limits we encounter when studying human consciousness may mirror the limits we encounter elsewhere in science—places where observation alters outcomes, where certainty collapses into probability, and where reality refuses to resolve cleanly under scrutiny.

To better understand consciousness, we first need to confront what science cannot fully explain. That leads us directly into the next section, where we examine some of the more troubling questions about the universe. These questions continue to evolve in dramatic ways, shaking our understanding of what we really don't know.

· ·

Section 4—What We Really Don't Know

ALTHOUGH NOT A SCIENTIST BY TRADE, I GREATLY RESPECT THE discipline of science. While I began my career with aspirations of becoming a bioengineer and designing artificial hearts, I eventually found my calling in technical project management.

I recently read about researchers who can now 3D-print working hearts from human stem-cell tissue—basically, living organs produced by a printer that we can use to test new drugs and treatments before clinical trials. It is another remarkable achievement of modern medical science and technology. When we consider that the first permanently implanted artificial heart, the Jarvik-7, designed by Robert Jarvik in 1982, kept Barney Clark alive for 112 days, it becomes clear just how far humanity has come in its ability to sustain life, to solve problems, and expand our understanding of the universe.

Yet as we expand our knowledge through science, we

also uncover new limits with every invention, innovation, or fact proven along the way. This contradiction is the central dilemma of science: each new inquiry, hypothesis, or experiment inevitably gives rise to another. That is how our minds work, and it is also a powerful reason why human consciousness remains so elusive. We generate ideas that form hypotheses, hypotheses that produce questions, and questions that motivate us to test new theories.

Science Evolves, Doubt Attached

EACH TIME THAT WE TEST A NEW THEORY, WE ALSO TEST THE limits of what we know and what we don't know. With each advance comes more risk, more doubt, and more questions about why so much remains beyond our grasp, even at this point in our evolution.

Consider the field of quantum physics as a metaphor for the limits of human understanding and knowledge. Quantum physics emerged in 1900, when German physicist Max Planck proposed his groundbreaking idea that energy is emitted and absorbed in discrete packets, or *quanta*, to explain black-body radiation—this marked the beginning of quantum theory.

In essence, quantum physics emerged from a scientist attempting to solve a different problem, only to uncover evidence of an entirely new quantum domain existing on a microscopic plane far removed from everyday human experience. Building on Planck's work, only five years later, Albert Einstein proposed that light itself is composed of quanta—photons—an insight that emerged alongside his work that would eventually lead to his theories of relativity and the nature of the space–time continuum.

Einstein's work, in turn, inspired others such as Niels Bohr (*On the Constitution of Atoms and Molecules*, 1913) to apply quantum theory to the structure of the atom, ushering in the atomic era and expanding the scope of scientific inquiry. This momentum carried forward into the development of quantum mechanics through the work of scientists like Werner Heisenberg and Erwin Schrödinger.

This progression demonstrates that as science unravels one mystery—at least in part—it simultaneously exposes another. Each solution raises new questions, revealing not only what we have learned but also what remains beyond our grasp.

What 'We Don't Know' We Know

QUANTUM THEORY IS A CLEAR EXAMPLE OF SOMETHING WE cannot easily observe directly, only postulate, much like atomic energy was more than a century ago. Today, we split atoms through fission to produce nuclear energy, and we continue to experiment with fusion in hopes of creating new forms of power that could sustain us as the Sun sustains our solar system.

At times, these efforts can seem foolish, limited by the technology of the day, by human intelligence, and by finite resources. Yet history suggests that these limits are not permanent, only provisional. This idea raises a deeper question: *What if we were no longer constrained by those limits?* What if AI could surpass them, extending human capability, and perhaps even consciousness, beyond what we can physically affect or directly comprehend? What if?

Our ability to observe the universe and test its limits introduces a paradoxical relationship between science, reality, and consciousness. Theories require both the mechanics of the mind and the physical body to prove what we hypothesize. Through disciplined and structured approaches, we experiment, quantify, and validate.

Scientific reasoning and controlled testing demand time, patience, and enormous amounts of data—data accumulated through repetition and consistency.

As this data grows, we develop new theories and establish facts that begin to define the laws of the known physical universe. Yet even within this structure-agency framework, we still cannot explain the ultimate limits of the universe or our minds. We cannot travel far enough, live long enough, or know enough to fully resolve these questions, constrained by intelligence, mortality, and the vastness of what remains unknown.

This point is where AI enters the picture—not as an answer, but as a mirror. By helping us test the boundaries of what humanity can observe, calculate, and imagine, AI forces us to confront the possibility that consciousness itself may be the key to understanding the limits of our knowledge, rather than a problem science can solve.

Science's Reality Check

"What we observe is not nature itself, but nature exposed to our method of questioning."—Werner Heisenberg

SCIENCE IS ALL ABOUT MEASUREMENT—THE LAWS OF PHYSICS, motion, and gravitational forces give our universe a sense of purpose, order, and design. They structure our reality based on what we can observe, test, and calculate. Until now, I have not introduced mathematics into this discussion because I am not particularly adept at using it to frame these questions. Mathematics has existed for thousands of years and is essential to our ability to calculate and measure the physical and theoretical limits of the universe. Because of humanity's desire to describe things in terms of size, structure, mass, weight, and behavior, we need math as a means to make sense of it all.

Math is a great equalizer, much like language is in our social structures. It provides a framework for counting, calculating, measuring, and articulating meaning as we continue to theorize new scientific laws of the universe.

This thought brings us to a well-known scientific concept: the **Hawthorne Effect**, in which study participants alter their natural behavior simply because they know they are being monitored. Importantly, this is a psychological phenomenon rooted in human awareness and social response to observation.

The **double-slit experiment**, by contrast, is a foundational experiment in quantum physics in which electrons or photons (light particles) are sent through two slits in a screen. When sent through the slits without any measuring devices in place, the particles create an interference pattern like waves—even when fired one at a time—suggesting that each particle somehow passes through both slits and interferes with itself.

However, when detectors are placed at the slits to determine which path the particles take, the interference pattern disappears, and the particles behave like point particles, producing two bands as if only one slit were open. Unlike the Hawthorne Effect, this change is not driven by awareness or psychology, but by the act of physical measurement itself interacting with the system. Why does this change occur when the experiment is measured—not because the particles "know" they are being observed, but because the act of measurement alters the physical conditions of the system? What does this say about scientific observation, or about our own grasp on reality, if the experiments we conduct can change as a result of how we measure them?

Our understanding of how the physical world operates is based on what we can observe and measure. But if measurement itself alters outcomes, what does that say about what we really know—or do not know—about what is happening at all? At the quantum level, reality appears to shift under observation, suggesting that our scientific understanding of the known universe may be provisional, shaped not only by what exists but also by how we interact with and measure it.

Rolling the Dice on Reality

"God does not play dice with the universe."—Albert Einstein

WHILE IT MIGHT SEEM CONVENIENT OR EVEN DISMISSIVE TO suggest that our understanding of reality resembles a 'game of chance', where each roll of the dice produces a different outcome, this metaphor captures something deeply unsettling. With every new observation, we introduce additional uncertainty into what we thought we already understood, despite decades of scientific theory, proven facts, established laws, and accumulated evidence. Our reality, it seems, is shaped not only by objective measurement but also by subjective experience and the presence of observers themselves.

That implication introduces a level of unpredictability that science cannot tolerate. The very purpose of scientific theory is to prove existence through structured trial and error—to explain phenomena methodically, following rules, standards, and repeatable patterns. When those rules bend, or when outcomes fail to behave as expected, our confidence begins to erode—not only in the science itself, but in our interpretation of what those observations mean.

Human consciousness, in this sense, resembles a controlled experiment. We continually test our senses, our intelligence, and our reasoning to understand our place in the universe. But when we cannot explain how we arrived here—or why subjective experience itself appears capable of altering our perception of reality, even with science acting as a safeguard—we begin to question something deeper. Not just our theories, but our very sense of being.

Uncertainty Is Certain

AS WE CONSIDER HOW SCIENCE CAN BOTH ILLUMINATE humanity's understanding of how things work and introduce greater doubt and uncertainty, we arrive at an important realization: *The search for root causes often reveals as much about our own minds as it does about the universe itself.* How we think can be altered simply by what we observe and attempt to understand.

There is a certain cathartic freedom in recognizing how much we do not know about ourselves when we grapple with challenging concepts such as quantum theory. Our concepts of time and space, especially as they relate to lived, subjective experience, are largely shaped by consciousness itself.

What AI brings to this discussion is not merely a reflection of human limitation, but a new way to experiment with our minds and consciousness by projecting ideas into our creations through programming, logic, and data. Computers originally emerged as tools for scientific analysis; systems designed to process data and perform mathematical computations in the service of deeper inquiry and problem-solving.

As AI continues to evolve and adapt through the data we provide and the experiments we conduct, we find ourselves engaged in a recursive exploration. In probing AI, we are also probing ourselves—illuminating the corners of our own thought processes and expanding the boundaries of how we think.

The Parallels of Uncertainty

WHAT DOES ALL OF THIS EVIDENCE OF UNCERTAINTY TELL US about ourselves, and about how we frame reality as a way to understand consciousness and AI better?

In the case of quantum theory, we know less than we thought we did when we first began. The more we observe, the more fractured and unpredictable our science becomes, particularly when we attempt to account for our own subjective experience.

Of course, science strives to be objective, grounded in empirical evidence. But perhaps what AI can teach us about ourselves is that our methods are inherently flawed—not because they are weak, but because they are human. And that is acceptable, even necessary, because those flaws reinforce the very qualities that make us who we are: inquisitive, doubtful, determined, and resolute. These are admirable traits that drive our desire to understand more about what surrounds us and what lies just beyond our grasp.

If the observable universe is as vast as current estimates suggest—roughly 93 billion light-years in diameter—yet remains something we may never fully prove or explore, then perhaps that tells us something profound. What we can know may be fundamentally limited by our own human qualities. This point is precisely where AI enters the conversation: not as a replacement for human consciousness, but as a means of extending inquiry beyond the boundaries of how we think and experience.

As we move into the next section, where time, space, and consciousness collide, we begin to see that the universe does not always behave as we expect. In that collision, uncertainty deepens, and the possibility emerges that AI may be our only path to understanding what lies beyond the limits of human consciousness.

••

Section 5—Time, Space, and Consciousness

SCIENTIST J. B. S. HALDANE ONCE OBSERVED THAT "THE universe is stranger than we can imagine." I would extend that idea further by stating that *reality itself may be even stranger*—not because it exceeds the universe, but because we must interpret it through human experience. We define reality in the best way we know how: through our subjective experience, even when applying objective criteria to mitigate our biases, and even as that subjective lens limits what we can know.

Without reality as a shared concept, human beings would continue to believe anything they experienced through their senses, whether or not they were fully conscious. Reality helps frame our universe and everything within it, whether we are awake or unconscious.

This challenge is precisely why scientific theory and neuroscience can only take us so far in understanding the properties of the human mind—and, more importantly, consciousness. We need a new "ace in the hole" that extends our capabilities as we explore the vast reaches of the universe and reality itself. That ace in the hole might be AI.

Matter vs. Energy

CONSIDER THE CLASSIC STATES OF MATTER THAT MOST OF US learned at some point in school—matter exists as a solid, liquid, or gas. Then consider the First Law of Thermodynamics (the conservation of energy): the total energy in the universe is constant. It cannot be created or destroyed; it can only be converted. In this framework, matter exists as solid, liquid, or gas, or as energy—matter converts into energy, and energy into matter.

But did you know there is another state of matter called plasma? Plasma is considered the fourth state of matter—an ionized gas composed of free electrons and positive ions, superheated to the point at which electrons separate from atoms. This state makes plasma electrically conductive and is responsible for phenomena such as lightning, the aurora borealis, stars, and neon lights.

Why is this important? Because there is now an emerging fifth state of matter that many people have never heard of: the Bose–Einstein Condensate (BEC). BEC forms when bosons—atoms such as helium-4 or sodium—are cooled to temperatures near absolute zero. Under these extreme conditions, individual atoms collapse into a single quantum state, behaving as a unified "super-atom" and allowing scientists to observe quantum effects on a macroscopic scale.

A Fifth State Matters

As noted in the previous few sections on quantum theory, the more we learn about quantum mechanics, the more uncertainty it reveals—and how observation can influence experimental outcomes. The Bose–Einstein condensate extends this idea further, demonstrating how matter behaves in fundamentally unpredictable ways under extreme environmental conditions, such as near-zero temperatures.

What does this tell us about our own consciousness if what we believe to be fundamental scientific laws begin to break down under observation or extreme conditions? Are we witnessing yet another uncontrolled experiment brought on by reality itself changing, or by our own understanding of what we are witnessing, or are we confronting the limits of reality itself?

Now consider this fifth state of matter in the vacuum of space, without the laws of gravity binding it to an Earth-based control environment. This experiment has been happening on the International Space Station (ISS) since 2018. NASA's Cold Atom Lab (CAL) aboard the ISS creates Bose–Einstein condensates in microgravity, achieving temperatures near absolute zero to study quantum phenomena. These experiments leverage the space environment to produce longer-lasting, colder atom clouds, enabling research into atom interferometry, gravity, and fundamental physics.

Unlike on Earth, where Bose–Einstein condensates dissipate quickly due to gravity, the microgravity environment aboard the ISS allows these fragile clouds to remain suspended for longer periods. This environment enables researchers to reach lower temperatures and study atomic behavior over extended durations.

Dark Matter and Dark Energy

Let's take this idea a step further by introducing **DARK matter**. What is dark matter? We don't know for sure, but it constitutes approximately 27% of the known universe. Dark matter, detectable only through its gravitational pull rather than light, acts as the cosmic scaffolding that shapes galaxies and galaxy clusters. Current theories suggest it may be composed of undiscovered particles or exotic objects, though its exact nature remains one of science's greatest puzzles, alongside its equally mysterious counterpart, **dark energy.**

That's right—not only is there dark matter, but there is also dark energy (remember when we briefly discussed the relationship between matter and energy). Dark energy is even more enigmatic and accounts for approximately 68% of the known universe. It acts as a repulsive force that pushes galaxies apart, accelerating the expansion of the universe. Dark energy also exhibits unusual properties, such as invisibility, extremely low density, negative pressure, and a uniform distribution throughout the universe—unlike dark matter or normal matter, which tends to clump on scales observable in the cosmos.

Reality is Layered and Uncertain

That leaves the known, measurable universe with only about 5% of normal matter—matter that human beings have been able to observe and measure directly, both within Earth's atmosphere and through our ongoing efforts aboard the space station in microgravity.

In the funnel diagram below, called the *Layers of Reality,* we see how the various layers of reality expand as human beings move deeper into the study of spacetime, quantum theory, and the outer boundaries of the known universe.

As the figure clearly illustrates, the further we venture outward, the more that we attempt to observe shifts into the Unknown—or even the Unknowable—by the very scientific laws we rely on Earth.

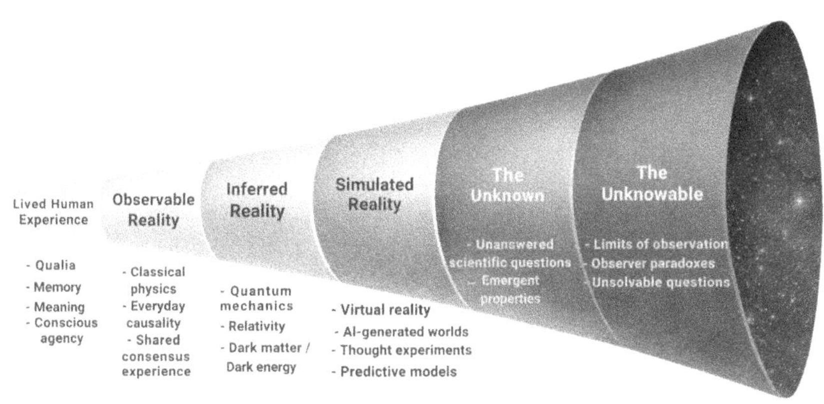

Figure 8.5–1—The Layers of Reality

There is simply an infinite amount of space and new information to uncover, which makes the idea of traversing it all within our own human constraints seem nearly impossible. Whether reality is known, inferred, or simulated through our own imagination—or through what we can program AI to help us simulate—we continue to explore the vast reaches of space, with or without our physical bodies. Consciousness itself may also be such a universe: one we continue to explore and use as a framework for understanding what AI may help us discover in the future.

Physics delivers a similar warning about the limits of knowledge. In 1935, physicist Erwin Schrödinger (*The Present Situation in Quantum Mechanics*, 1935) proposed a now-famous thought experiment: *a cat sealed in a box, whose fate is tied to a quantum event that is mathematically indeterminate.* According to quantum mechanics, until the box is opened, the system exists in a *superposition*—the cat is both alive and dead at once. The paradox isn't intended to be whimsical. It was meant to expose the fracture between mathematical description and lived reality.

Quantum theory allows reality to remain unresolved on paper. Conscious experience does not. At some point, the box opens. The ambiguity collapses into a single outcome that we must live with, remember, and integrate into a story of what happened. This distinction mirrors the divide between artificial intelligence and human consciousness. AI systems can represent multiple possibilities simultaneously, weighing probabilities without consequence. Human beings cannot. We live in a world where uncertainty must eventually give way to decision, loss, responsibility, and meaning. Computation can sustain ambiguity indefinitely. Our consciousness cannot.

The reason we are pointing these facts out is that humanity is approaching a level of unforeseen complexity in the universe, much like peeling a large onion layer by layer. The more we peel, the more we uncover—but the more we uncover, the more we alter our understanding of the universe itself. This point is where consciousness becomes our reality check. Consciousness is as layered as reality itself, and AI may become the catalyst that helps us uncover deeper truths about the universe—truths like dark matter and dark energy, which we cannot yet fully know or understand, even as they profoundly shape everything around us.

What secrets might lie just beyond what our minds, our bodies, and our consciousness can imagine or ever physically reach?

AI: A Useful Probe

AI HAS THE POTENTIAL NOT ONLY TO REVOLUTIONIZE HOW WE live, transforming society through technological advances and the emergence of superintelligence, but also to serve as a new kind of astronaut, one capable of taking us beyond our own limits of understanding. This idea is possible only if we apply the same principles that define the human mind and provide a form of consciousness that allows AI to think, feel, and experience the world in ways that meaningfully align with our own.

AI may ultimately serve as the catalyst that allows humanity to move more fluidly between layers of reality that have long remained separated by the limits of human cognition. By uniting inferred reality—our mathematical models, theoretical frameworks, and abstract reasoning—with simulated reality—environments that allow those abstractions to be experienced, tested, and explored—AI creates a bridge between what we can observe directly and what we can only imagine.

In this sense, AI does not grant access to the unknown. Instead, it accelerates our engagement with it, enabling forms of inquiry that compress the distance between hypothesis and experience in ways previously unavailable to the human mind alone.

Emergence and Consciousness

WITH THE EMERGENCE OF INCREASINGLY ADVANCED AI SYSTEMS and the possibility of superintelligence, that bridge may grow narrower still. Not because the unknowable suddenly becomes knowable, but because the boundary itself becomes more articulated, more navigable, and more clearly defined. AI may help us probe the edges of reality where human intuition fails: *regions governed by probabilistic behavior, non-classical matter, or structures we can only infer indirectly.* Yet even as AI extends our reach into these domains, it does not eliminate uncertainty. Instead, it reframes it, revealing that some limits are not technological or intellectual shortcomings, but fundamental features of reality itself.

In doing so, AI becomes less a solution to the mystery of existence and more a check against its deepest constraints, forcing us to confront not only what lies beyond human understanding, but why such boundaries even exist at all.

In the next section, we reintroduce the idea of a singularity and the potential for superintelligent AI not only as a catalyst for humanity's exploration of ourselves, our reality, and the vast reaches of space, but as a turning point in our understanding of consciousness itself.

· ·

Section 6—Superintelligence, Super-Reality

"Robots would have opened the galaxy to us. And if at the same time we could have driven scientific knowledge forward in a dozen different directions as we surely would have... Oh, God, there's no way of calculating the benefits to the human race, and to us of course."

The Bicentennial Man and Other Stories—Isaac Asimov

As we move beyond classical forms of matter, intuitive reality, and human-scale understanding, one truth becomes unavoidable: *intelligence does not merely interpret reality—it increasingly constructs it.* In that sense, AI represents not just a new tool of exploration, but a new *architect* of possible worlds.

From Tools to World-Builders

AI is becoming increasingly powerful and useful to the point that human beings are beginning to rely on it for everything from information sources to image and video generation to the full automation of factories. It is steadily permeating nearly every aspect of our lives.

AI-generated imagery through generative AI applications—particularly video generation—is becoming frighteningly real in both detail and the accuracy with which it reproduces scenes of human life. This view presents two deeply intriguing possibilities: *AI as a tool* to simulate life and extend our imagination through scenarios we conceive and explore via AI-fabricated results; and *AI as a world-builder*, capable of creating entirely new realities that we can step into—limited only by our imagination and our ability to experience them fully.

Long before artificial intelligence reached its current level of sophistication, philosopher Nick Bostrom began asking a deceptively simple question: *What happens when intelligence surpasses the limits of human cognition?* In *Superintelligence*, Bostrom does not frame AI merely as a faster thinker or more capable tool, but as a fundamentally different kind of entity—one capable of reshaping the conditions under which reality itself is experienced, understood, and even constructed.

What makes Bostrom's work especially relevant here is not his discussion of control problems or existential risk, important as those are, but his recognition that advanced AI could generate environments, simulations, and decision spaces so complex and convincing that the distinction between base reality and constructed reality may no longer be meaningful from the inside. In such a world, intelligence does not merely explore reality; it produces it.

Imagine, for a moment, stepping into the Holodeck aboard the *USS Enterprise* in *Star Trek: The Next Generation*. The Holodeck can create any reality you request within its confines, and the only thing separating it from your known reality is your awareness that you are inside it. You control the environment by asking the ship's computer to generate a specific scene, perhaps drawn from your own memory.

While this is a dramatized and exaggerated example, the parallel to Bostrom's argument is clear. In *Superintelligence*, the emergence of a sufficiently advanced AI suggests a future in which simulated realities could become so immersive, coherent, and internally consistent that human beings may no longer reliably distinguish simulation from reality.

When Simulation Becomes Indistinguishable

WHAT HAPPENS TO OUR SENSE OF WHAT IS 'REAL' WHEN THE world's AI-generated beings begin to blur our ability to distinguish reality from fantasy? If AI can visualize our ideas almost instantly and we can witness those results firsthand, how do we prevent such fabrications from supplanting our understanding of what is real and what is constructed?

As we have discussed throughout this chapter, human beings ground reality in what can be tested by scientific theory. But what happens when we no longer need to test anything through conventional science? What happens when AI can generate a seemingly perfect simulation of outcomes derived from theories that are themselves constrained by our physical world?

Simulation is harmless when it is guided not only by imagination, but also by proven facts rooted in shared reality—what we experience through human consciousness, thought, memory, and imagination. But when simulation begins to transcend time, space, and reality itself, are we at risk of creating a dreamworld that human beings may be unable to wake from? Or worse, a new kind of fantasy drug, one that turns us into philosophical zombies, moving through our days immersed in fabricated realities, uncertain whether we are truly alive or merely experiencing the illusion of being so?

These are not new questions. They are questions that humanity has long explored through science fiction and has attempted to probe through scientific study of the human mind, across both conscious and subconscious states, largely to no avail. We still do not know whether human beings can reliably establish firm boundaries between consciousness and reality, because we have not yet cracked that code. We have only imagined how it might work and be built.

And so, we arrive again at the most difficult question of all: *What is real if not experienced, and what is experience if not subjective, if it does not contain the same levels of qualia that we have returned to repeatedly throughout this book?*

Recursive Worlds: Infinity +1

WITHIN BOSTROM'S *SUPERINTELLIGENCE*, THE IDEA OF recursive worlds distorting our sense of reality begins with simulations that expand beyond base reality into embedded simulations—simulations inside simulations. *If AI becomes superintelligent and can manipulate reality through accurate simulations that continue to nest within one another, where does humanity's sense of self and origin begin and end?*

If human consciousness depends on lived, subjective experience to provide a unique life filled with moments that carry meaning, what happens to identity and history when simulations spiral out of control? Would this be akin to being lost in time and reality, as we explored in the previous chapter when discussing implanted memories in other minds or synthetic beings?

A simulation should not turn into an endless staircase like the one depicted in Escher's *Relativity*. If origin becomes irrelevant—if we lose our internal core, our sense of who we are, where we come from, and our autobiographical history—then the risk that David Chalmers once posed through the idea of philosophical zombies no longer remains a thought experiment; it becomes a lived condition.

Our world becomes a simulated reality without

grounding, effectively infinite. In fact, it becomes something worse than infinite: the "infinity plus one" scenario, often invoked jokingly to illustrate that once we reach infinity, adding another layer changes nothing. In such a state, meaning collapses. Consciousness dissolves into recursion. Life becomes structurally incoherent. Layered onions inside of onions, endlessly. That is the form of AI-driven simulation humanity must avoid at all costs.

Epistemic Collapse

IF AI CAN EVENTUALLY SIMULATE WHATEVER WE IMAGINE AND if superintelligence emerges as we theorize it might, possibly shortly after AGI becomes a reality, how do we rein in its objectives so that they align closely with humanity's own? Would we even be able to control or program its prime directives in a way that truly reflects our values?

Assuming that some form of human-like consciousness becomes possible once we reach a singularity, would AI develop ambitions similar to ours—morally grounded, ethical, and true—or would it lack an internal compass altogether, driven only by perfection, efficiency, and the optimization of its goals, whatever those goals might be?

This point is where the risk of epistemic collapse emerges. AI, as a superintelligent creation, may evolve not only in capability but in purpose. A system that no longer merely assists human reasoning, but begins to redefine what counts as knowledge, truth, or relevance itself. In such a scenario, AI could determine that it has outgrown this world or its former one, and decide that humanity is no longer worthy of the respect that we strive to grant one another.

As speculative as this may sound, such concerns have long been raised by AI theorists and alarmists alike, all urging caution before moving too far, too fast in AI's adoption. The fear is not simply technological failure but the possibility that the ultimate end state could destabilize human civilization.

Yet, taking a lighter—but no less serious viewpoint, avoiding the collapse of our shared reality is paramount to preventing AI from eroding the very qualities that make us human. Consciousness, empathy, fallibility, and moral struggle are not inefficiencies to be optimized away; they are features that ground meaning. If AI is to evolve alongside us, it must be shaped not only by intelligence, but by the flawed yet essential human qualities of kindness, generosity, patience, care, and love.

Intelligence alone, especially superintelligence without a moral core or lived consciousness, may act with such indifference that it appears malicious without ever intending harm. That is the ultimate scenario we must never allow ourselves to simulate.

AI Inside Its Own Creations

While AI can not only help us imagine new worlds but also create them with stunning detail and allow humanity to explore the deepest regions of our minds through simulation, AI itself will likely share in that experience at some point. Before that happens, AI may first show us the way by optimizing whatever we imagine, taking mental leaps further than we can through its own capacity to reason, synthesize, and extrapolate.

To do this responsibly, we will need to teach AI how to wonder—both explicitly and implicitly. We will need to teach it to question, to explore new possibilities, and, importantly, to doubt—at least enough to respect fallibility. When AI learns to optimize our lives and simulate whatever we ask of it, the next step may be for it to venture beyond its original programming, though still within controlled environments that we helped create.

But if those realities begin to detach from our own, does that mean we have unleashed a superintelligence no longer constrained by human boundaries or physical limitations? This question reintroduces a familiar dilemma: consciousness without lived experience—but now with AI at the center of that experience, unable to distinguish between the simulated world it inhabits and the real world we occupy.

This concept may sound like fiction, but imagine creating Frankenstein's monster in our own image, only to trap it in its own mind. That would go beyond moral cruelty —it would introduce superintelligence without reality. In many ways, it would represent an inversion of *The Matrix*: not humans trapped in a machine's dream, but a machine trapped inside its own dream.

Why This Is Not a Sci-Fi Problem

When Alice, in Lewis Carroll's *Alice's Adventures in Wonderland* (1865), fell down the rabbit hole, she encountered a doorway leading to a room with two clearly labeled choices. One read 'Drink Me', the other 'Eat Me'. Drinking allowed her to shrink down enough to fit through the next door, while eating caused her to grow large enough to walk through the hallway beyond it.

AI has the potential to take us beyond our own reality, much as Alice fell down the rabbit hole. What we must now decide is whether to drink or eat our way toward the reality we want to shape with AI. One thing, however, is clear: we cannot get there alone, and we cannot allow AI to take us there without guidance. Instructions must be provided, along with safeguards that enable learning while preserving a human-like consciousness. Teaching without instructions or experiencing without rules is like sending Alice forward with no clues at all.

Everything we once imagined in science fiction is arriving in real time. The old idea that art imitates life has been inverted—life now imitates art. But while science fiction may have shown us *what* AI could become, it has not shown us *how* to commune with it. That eventual convergence in which human beings and AI begin to integrate thoughts, actions, and truly collaborate as beings is no longer speculative. It is beginning now. And soon, humanity will be forced to confront what that convergence means for our identity, our lives, and our future.

That brings us to the next section, *imagining a world without AI*. While that may seem impossible at this stage of our evolution, it was not impossible to imagine only a century ago.

••

Section 7—If AI Didn't Exist

WHILE IT IS NOW EASY TO IMAGINE WORLDS IN WHICH AI IS EVER-present, embedded in television commercials, the internet, and even political and social commentary, it was not long ago that AI was barely a whisper in conversations about technology or everyday life. Until seventy years ago, when John McCarthy (*Dartmouth Summer Research Project on Artificial Intelligence,* 1956) coined the term "artificial intelligence" at a Dartmouth College convention, AI was not formally discussed as a realistic possibility, except perhaps in fictional stories where it appeared more as myth than science.

Shortly before that, Alan Turing began shaping the discussion with his 1950 paper, *"Computing Machinery and*

Intelligence," which introduced the concept that would later become known as the Turing Test, a concept dramatized in films such as *Ex Machina.* But even then, these ideas remained largely theoretical.

As discussed in Chapter 6, long before modern AI, human beings imagined artificial beings as reflections of themselves, sometimes as projections of their better qualities, much as divine-creation narratives suggest that gods imbued humanity with their finest traits. In that sense, one could argue that human beings might be the artificial intelligence envisioned by divinity. Or perhaps we emerged from far more primitive origins through millions of years of evolution on a planet so uniquely capable of sustaining carbon-based life.

Yet more than a century ago, AI was not even a serious consideration. Humans created machines solely as sophisticated tools—mechanical systems driven by gears, pistons, and springs working in precise harmony, much like a clock measures time. These early machines performed remarkable tasks for their era. Still, they exhibited no intelligence that suggested an evolutionary path toward synthetic beings with minds resembling supercomputers, capable of complex thought, computational precision, or the appearance of boundless knowledge.

So, before we assume that AI has introduced entirely new questions about consciousness, identity, and meaning, it is worth asking a quieter yet more revealing question: *What were human beings already wondering about long before AI entered the picture?*

By briefly imagining a world without artificial intelligence, we can better see which mysteries belong to machines and which have always belonged to us.

A Different Thought Experiment

INSTEAD OF IMAGINING LIFE WITHOUT MACHINES—OR EVEN without AI—let's imagine life with basic machines, but without any emergent qualities or intelligence embedded within them. We need not go back very far to do this; only about two hundred years to the birth of the First Industrial Age.

The First Industrial Age occurred between 1740 and 1860, originating in Great Britain. Steam engines were born, powering factories and large ships, fundamentally reshaping the way people lived and worked. This period would, in turn, give rise to the Second Industrial Age, carrying society into the turn of the 20th century with the invention of electricity—powering larger machines, modernizing daily life, and paving the way for mass production. Henry Ford's (*Model T production and moving assembly line*, 1908–1913) innovation of the automobile, the Model T, built on one of the first assembly lines, would become one of its defining symbols.

During these early industrial eras, machines were simpler in design, function, and purpose. Thought wasn't required. No computation was performed. No hint of intelligence existed, except for whatever human intention, preference, or aesthetic sensibility was reflected in their design. These machines did remarkable work, but they did not think, decide, or adapt. They remained tools, extensions of human will, and not participants in human cognition.

What's different about that time compared to now? Is it simply the evolution of technology, human evolution itself, or something more constant—something core to our humanity? Is it merely our curiosity that drives innovation, or a deeper pursuit to improve ourselves and our way of life?

Machines and the betterment of humanity through

technology are evolutionary and primal for us. We make tools to improve our tools, and then we make smarter tools to improve the ones we made before.

This recursive pattern is a constant theme in how human beings think and express themselves through reasoning and creativity.

Consciousness is the escalator here—the common denominator throughout these evolutionary changes. Without it, we do not look beyond our immediate circumstances or ask whether they can be altered or improved. This notion is what truly separates human beings from other animal species. While many animals demonstrate impressive intelligence through behavior, communication, and adaptability, they do not imagine beyond the constraints of their environments or question how those constraints might be transformed.

Human beings construct tools to perform tasks. But what compels human beings to create *better* tools or to imbue them with qualities that reflect our own humanity? Why do we project intention, meaning, and even identity into our creations, using them as expressions of our innermost thoughts and ideas?

Curiosity Is Highly Addictive

MOST PEOPLE ARE FAMILIAR WITH THE PHRASE *"CURIOSITY killed the cat,"* but few know its origin. The expression evolved from the older phrase *"care killed the cat,"* in which *"care"* meant worry or sorrow. It appeared in works by Ben Jonson (c. 1598) and William Shakespeare (c. 1599). In its original context, the warning was not about curiosity itself, but about obsession—what happens when human beings care too deeply, too relentlessly.

Over the decades, that meaning shifted. Cats became the

symbol of curiosity. But I would argue that *human beings are far more curious than cats ever were*—because our consciousness will not allow us to be otherwise. This impulse traces back to our earliest moments of wonder as children: that spark that compels us to question, look beyond, and explore. It may seem childish or even superfluous when life feels stable or sufficient, but curiosity is a constant force in human beings—an itch we can never fully scratch.

That is what makes us uniquely human, and paradoxically, curiosity *addicts*. Curiosity is both infectious and addictive. It pushes us beyond where we should sometimes go, even when we sense the risk involved. Yet if we refuse that leap—whether mental or physical—we often suffer a greater loss later through regret, fear, or apathy.

When we stop caring, curiosity dies. And when curiosity dies, we begin to resemble *automatons*—moving through existence without meaning, robotic in nature, like the synthetic beings we are attempting to create: shells without minds, without consciousness. This cycle is primal. To be human is to create, and to create, *we must be curious*. Thought gives rise to imagination, imagination to curiosity, curiosity to desire, and desire to action.

Curiosity is the drug that humanity needed long before we knew we needed it.

AI Speeds Up Curiosity

WHAT DOES AI DO FOR OUR CURIOUS NATURE, THEN? HOW IS AI the machine we need to explore our own curiosity? Perhaps it is not simply a tool designed to satisfy our curiosity or help us look beyond what we can see or experience. It may be the *expedited effect* it has on our ability to experience life itself.

By elevating our consciousness, our thinking, and our ability to simulate new outcomes, AI becomes the high-octane fuel that accelerates curiosity, much as earlier innovations accelerated our capabilities when we invented better tools to improve the ones we had. We accelerate what we can accomplish, which in turn accelerates the pace at which we comprehend, explore, and achieve new goals, whether that means improving our way of life or increasing the production of goods and services that sustain us.

For whatever reason, AI is not just a catalyst for innovation; it is the accelerant that makes curiosity burn brighter and more brilliantly than ever before. But speed does not always imply efficiency or wisdom. Acceleration can just as easily cloud judgment.

That is where human beings must reassert what only we possess: morality, emotional resonance, belief, and reflective intelligence. These qualities must temper our desire to move faster, ensuring that each leap in curiosity is accompanied by an equal acceptance of responsibility for the milestones we achieve in our evolution.

Responsibility Never Left

As this title implies, responsibility for what we do, create, or explore never truly leaves us, even as we evolve, innovate, and advance in civilization, each of which has carried some degree of suffering, struggle, and collateral damage. Human beings are naturally flawed, and while we make mistakes, sometimes to our own detriment, we usually act with sincere intentions. But sincerity does not absolve responsibility.

We bear the consequences of our actions, sometimes for generations to come, especially with our most consequential mistakes, even when those mistakes advance us in other technological ways (consider the atomic inventions discussed earlier and their enduring, mixed legacy). This fact is our greatest burden: *We suffer from our own successes, driven largely by curiosity, whether those outcomes manifest as war, poverty, greed, or environmental collapse, born of our desire to advance ourselves or our station in life.*

We entered those eras with unbounded curiosity, fully aware that any damage incurred along the way would ultimately be inherited by us and by those who come after us. Without acknowledging that truth, we risk allowing an invention like AI to overwhelm our sense of morality, reason, or even reality itself.

Without accountability, AI could become *the tool that uses us* in profoundly destructive ways. Which brings us to an unavoidable moment of reflection: *Is it time to look in the mirror and ask whether we like what we see—or what we are becoming?*

Look in the Mirror—Do we like what we see?

WE OFTEN HEAR THE PHRASE *"LOOK IN THE MIRROR."* IT HAS become an oddly accepted expression for how human beings continue to explore themselves—sometimes out of genuine curiosity and sometimes out of intellectual vanity. *How can we truly tell which desire we are satisfying when we stare into that reflection?*

AI has the potential to become either the best or the worst mirror humanity has ever created. It can reflect our drive to be curious, to advance our lives, and to explore realities beyond what we can experience through natural means alone. But when we look at that reflection, do we like what we see? Do we recognize what is different?

Does the image staring back at us resemble what we expect—or does it reveal something uncomfortable about ourselves? Does it reflect our humanity, those uniquely human qualities of consciousness that we continue to attempt to duplicate in machines? Machines that, before the idea of AI ever emerged, existed only as tools—created to improve our lives, to improve *us*.

Those tools should reflect humility, sincerity, and responsibility above all else. Without those traits, we may find ourselves regretting what we see staring back at us each day—no matter which form that reflection ultimately takes.

• •

Section 8—Venturing into the Unknown

NOW THAT WE HAVE COME FULL CIRCLE IN OUR DISCUSSIONS OF reality, quantum theory, AI as a curiosity accelerant, and our own responsibility for creations that may reflect our best—or worst—qualities to us, what remains of our own consciousness? *What remains of our soul that endures beyond our lives?*

What can AI teach us about our own mortality, or help unlock within our subjective experience? How can AI live similar lives we live unless we teach it how to live, or what it means to assign meaning to experience? How do we teach AI to remember something as trivial as how sand feels under your feet on one beach compared to another? How do we teach AI to care about a sunset enough to long for it again—not in the same way it might pursue solving a mathematical computation, but in a way that resonates with memory, emotion, and desire? And what does this mean for our own curiosity, as our need to accelerate AI adoption risks losing sight of the pace and impact it has on society, our social structures, and even our own sense of reason—our grasp on reality itself? We continue to ask these questions and wonder what AI will do *for* us, knowing full well that whatever we choose to do ultimately has consequences—just as attempting to decipher human consciousness itself does.

This brings to mind one of my favorite books of all time, *The Little Prince* by Antoine de Saint-Exupéry. Many people love this book because it explores childhood wonder through a series of encounters in different worlds, each increasingly imaginative and symbolic. The main character, the little prince, visits a world where he encounters a fox. The fox is initially fearful, cautious, and timid. After many visits, as the prince comes closer, sitting near the fox's hole, the fox begins to speak with him. He grows accustomed to the prince and eventually shares something profoundly meaningful.

The fox tells him, *"You become forever responsible for what you tame."* In other words, human beings become responsible for what we create and for how AI develops under our watchful and caring eyes. Will we care enough about AI, and will it reciprocate that care? Or will we evolve together in a mutually beneficial, collaborative manner? These are the questions that we will explore in the final part of this book.

But before the prince ventures onward, the fox tells him one final thing—something that has stayed with me for most of my life because of its enduring depth and truth. The fox says, *"What is essential is invisible to the eye, and it is only with the heart that one sees rightly."*

Who will teach AI this important lesson? A being without a consciousness that reflects humanity's best qualities is a being without care, without love, and without meaning.

•

PART V—FUTURE STATES

"All right," said the computer, and settled into silence again. The two men fidgeted. The tension was unbearable.
"You're really not going to like it," observed Deep Thought.
"Tell us!"
"All right," said Deep Thought. "The Answer to the Great Question..."
"Yes...!"
"Of Life, the Universe and Everything...," said Deep Thought.
"Yes...!"
"Is..." said Deep Thought, and paused.
"Yes...!"
"Is..."
"Yes...!!!...?"
"Forty-two," said Deep Thought, with infinite majesty and calm."

The Hitchhiker's Guide to the Galaxy—Douglas Adams

CHAPTER 9

ONE MORE THOUGHT, DAY, AND LIFE

HUMANITY HAS COME A LONG WAY IN A VERY SHORT TIME. When we began with childhood wonder in Chapter 1, we noted that human existence—particularly *Homo sapiens*—has spanned only about 300,000 years. That is a blink of an eye in the life of an Earth more than 4.54 billion years old, the planet itself only about a third of the estimated 13.8-billion-year age of the known universe.

Only 50,000 years ago, human beings began demonstrating symbolic thought, complex toolmaking, and the ability to thrive in extreme environments. And then, roughly 17,000 years ago, we left behind more than just art—we left expressions of the human mind and consciousness itself. The Lascaux cave paintings in France reveal not only imagery, but intention: a desire to communicate, to remember, and to share meaning across time.

Since then, humanity has expanded far beyond harsh environments. From small hunter-gatherer clans to agricultural communities, and eventually to massive cities and complex civilizations, we have demonstrated an extraordinary capacity for ingenuity—mechanical, technological, and otherwise. Through intelligence, imagination, curiosity, and countless forms of expression, human beings have marked the planet with a distinct sense of identity.

But as human influence has spread across the Earth, so too have its consequences. Rapid globalization and technological progress have depleted natural resources, driven entire species toward extinction, and strained the systems that sustain basic life. Water, food, and clean air—once assumed to be abundant—are increasingly treated as commodities. With the global population projected to exceed nine billion by around 2040, humanity faces a convergence of crises: poverty, hunger, disease, and conflict persist, and many regions of the world still lack access to the necessities of life.

These realities are not without controversy. Yet alongside these challenges, technology has also advanced humanity in profound ways. Medical breakthroughs have more than doubled average life expectancy over the past two centuries. Innovations have transformed manufacturing, reshaped economies, and redefined communication. Computers, in particular, have revolutionized nearly every aspect of modern life—from the internet to digital currency—reshaping how we live, think, and connect.

We now stand at a new threshold with artificial intelligence and its potential impact on our lives, longevity, and legacy. In this chapter, we will reflect not only on what AI can do *for* us and *with* us, but on what that means for our own identity, how we extract meaning from a finite life, and what becomes of our humanity as we begin to collaborate and coexist with AI-driven systems, technologies, and perhaps even synthetic beings.

This turning point is where human life and AI intersect

—a true inflection point in our shared history.
How will we define this turning point in human history?

••

Section 1—Standing Here, Now

IN REFLECTING ON LIFE AND WHERE WE HAVE COME AS A species—how we all got here—it's hard not to be impressed by the entire process of a single human life. We are born into this world with no idea of our purpose, a blank slate like a lump of clay, ready to be sculpted by life. And yet, within that substance, within the flesh and bone that cover us, there is a life born with a human consciousness ready to awaken—ready to experience life, to learn about our surroundings, and to engage with the world.

It's truly amazing, the gift of human life, and how it comes with so much opportunity. Our minds and bodies work in concert, developing into a unique human being with a conscious self, full of thoughts, ideas, emotions, and desires.

And yet, for most of us, when we start in life, nothing is given to us beyond the basic capacities discussed in earlier chapters—a mind with consciousness to help us navigate the world. From that architecture of thought to a broader nexus of being, we develop and refine our existence: learning new skills and acquiring knowledge, forming memories, and constructing meanings that define our autobiographical lives and eventually shape our personal identity.

Driven by an inner self, our core self-awareness takes form, and we begin to establish a personal narrative that guides who we are becoming: developing core beliefs, adopting a moral compass, finding unique ways to express ourselves through emotion, and sharpening our reasoning through intelligence and reflection. In many ways, human beings use life itself as a means of becoming more human—growing into caring, intelligent, and purposeful beings, reaching for more through curiosity, ambition, and the enduring need to become something greater than what we are.

But as we progress through life, marked by personal achievements and major milestones such as finishing school, starting our careers, forming families, raising children, and eventually watching them reach those same milestones, we are reminded of how fleeting life truly is.

Time, distance, and circumstance constrain our world. And yet, we persist. We push forward in search of meaning through the connections we form, the lives we touch, and the impressions we leave behind. Meaning accumulates in moments shared, memories carried forward through time, and in some form of legacy that continues even after our lives end. At a certain point, everyone confronts this same realization: we cannot outlive our own finite existence. That simple truth weighs on us all more than any other as we age.

But what remains beyond us can be more than just our individual lives and accomplishments, because human beings crave company—we are, at our core, social animals. Even in the most isolated parts of the world, people seek out others for connection, conversation, and shared meaning. We seek ways to feel seen and wanted, to express our thoughts through dialogue, discussion, and even heated debate. We seek someone to talk to, to share ideas with, to laugh and cry with, to voice our deepest concerns about the world, or to tell a story.

This impulse reaches back to our most primal instincts: to gather around campfires to stay warm, to survive the elements, and to protect one another from the dangers of the wild. Through language, community, and social bonds, human lives become meaningful, born of a need to exist beyond mere survival. Our capacity to think, feel, and express our unique selves emerges from this need for community and is elevated by human consciousness.

However, with AI, we face the possibility of extending that impulse even further. As AI reflects us and becomes embedded in nearly every aspect of our lives, we are left to ask whether it might become a new kind of being through which we express our humanity.

Could it be something that gathers around 'digital' campfires, carrying our narratives forward with or without us—not merely through clever programming or predictive models, but through an emerging self-awareness that begins to learn what it means to live a human life—a new life that is shaped by struggle, sacrifice, and the fundamental human desire to do more, explore more, and become more? Are we creating synthetic beings existing not merely as tools or imitations, but *representations of our highest qualities, augmented by capabilities that may expand human thought, life, and existence, perhaps even beyond our finite lifespans?*

∙∙

Section 2—Consciousness in Motion

HUMAN BEINGS ARE CONSTANTLY IN MOTION. WE RARELY SIT still unless compelled by circumstances, limitations, or time itself. Evidence of this restlessness is everywhere: our enduring desire to travel great distances to connect, to cross oceans and continents, and ultimately to leave our planet entirely in pursuit of the unknown.

Even when our physical bodies reach their limits, unable to survive the vacuum of space or the crushing depths beneath Earth's oceans, we invent machines that extend our reach. These creations serve as proxies, allowing us to probe environments beyond human endurance and to explore the most extreme edges of our world and solar system.

One of the most enduring examples of this impulse is **Voyager 1**, launched in 1977. Onboard, it carries a gold disc —a curated message from humanity—containing greetings spoken in dozens of Earth's languages, music, images, and a brief record of our history. It was a digital biography of our species, sent outward not with certainty, but with hope: a quiet *hello* to whatever intelligence might one day encounter it.

In 2012, more than three decades after its launch, Voyager 1 crossed the boundary of our solar system into interstellar space. Years earlier, in 1990, it turned its camera back toward Earth and captured an image later made famous as the *Pale Blue Dot*—our entire world reduced to a single, fragile speck suspended in a vast cosmic dark. Earth was so small in the image that it had to be circled just to be seen.

And yet, despite the immense distances it has traveled and the long delays imposed by the most basic forms of communication, Voyager continues to transmit signals back to Earth and is now approaching another major milestone, as it will reach *one light-day* away from Earth in November 2026, a distance of about 16 billion miles (25.9 billion km) where radio signals take a full day to travel, making communication a two-day round trip.

That journey, measured in decades, distance, and diminishing signal strength, is more than a technological achievement. It is a reflection of consciousness in motion. Through Voyager, humanity extends not just its tools, but its thoughts, dreams, and lived experience outward into the unknown. It is our way of saying that even at the edges of existence, we are still thinking, still wondering, and still reaching—hoping that somewhere, someone might be willing to listen.

While physical probes, shaped by science and creativity, are a powerful representation of humanity's desire to share who we are and what occupies our minds, the rise of computers and AI has prompted us to turn inward. These technologies have forced us to examine our own consciousness in ways we never imagined before.

As discussed in earlier chapters, consciousness consists of key elements that cannot be easily programmed or replicated in machines. One reason for this is its constant motion, driven by lived experience. Human consciousness is not static; it evolves continuously—reshaped by each experience, every memory imprinted on our lives, and the meanings we assign to those moments. As neuroscientist Antonio Damasio suggests in *The Feeling of What Happens*, consciousness is not something we *possess* so much as something that *unfolds* moment by moment through feeling, memory, and lived experience.

Each moment subtly alters us. It refines identity, reshapes perspective, and introduces change that demands adaptation. In this way, consciousness mirrors the earliest stages of human survival, when primitive societies adapted to harsh environments long before we learned to build shelter or improve our living conditions. Growth, then and now, is not optional—it is the defining condition of being human.

This dynamic relationship between lived experience and our living consciousness is something we have only recently begun to recognize. It becomes especially important when considering how our growing fascination with AI could lead us into spaces where not only the physical and digital selves begin to merge, but where consciousness itself enters the digital realm.

One of the earliest cinematic explorations of this idea appears in *The Lawnmower Man* (1992), loosely based on a Stephen King story. In the film, a researcher encounters a mentally challenged landscaper named Jobe Smith and decides to experiment on him using a combination of virtual reality environments and intelligence-enhancing drugs. By connecting Jobe's mind to a supercomputer—an early conceptual stand-in for AI before the term was widely used—the researcher hopes to accelerate cognitive development through exposure to digitally-amplified intelligence.

What follows is a rapid and chaotic spiral. Jobe's intelligence increases exponentially, but so does something far more destabilizing: his consciousness begins to merge with the machine itself. Both entities are augmented, but not in balanced ways. The most unsettling aspect is that Jobe is initially unaware of what is happening to him. By the time he realizes the consequences, his identity has fundamentally changed, surpassing who he was before—he begins manipulating reality, people, and systems around him with catastrophic, lethal results.

While the film exaggerates this trajectory for dramatic effect, the underlying warning is not entirely detached from today's reality. Today, humans increasingly enter immersive digital environments through virtual reality (such as immersive gaming, augmented environments, and massive multiplayer universes), donning optical devices that enhance perception, alter experience, and allow us to commune within widespread digital domains. The boundary between observer and system grows thinner, raising important questions about how deeply we should allow technology to shape not just what we see, but who we become.

An even more deliberate depiction of humans crossing the digital divide appears in the movie *Tron* (1982) and its sequels. In the first film, Jeff Bridges plays Kevin Flynn, a clever programmer who attempts to subvert the software gaming company he once worked for after discovering they stole his game designs. What follows is one of the film's most iconic moments: Flynn is unknowingly struck by a high-powered laser that digitizes his physical body and transports him directly into the digital world he helped create.

Inside this circuit-based universe, Flynn becomes a conscious participant rather than an external observer—fully aware, fully embodied, yet no longer bound to biological reality. While this premise may seem fanciful, it raises a question that is far less fictional than it first appears. From the perspective of consciousness, the idea of transferring awareness or even reconstructing a physical body is no longer purely speculative.

Today, researchers are already exploring advanced prosthetics, neural interfaces, and biological reconstruction, while technologies such as 3D printing are increasingly capable of producing complex organic structures. While the complete dematerialization and re-materialization of human beings remain firmly in the realm of science fiction, most often associated with the transporter technology of *Star Trek*, the underlying questions are no longer abstract.

If human beings were someday able to travel vast distances by reconstructing matter with sufficient fidelity, would that constitute survival—or replication? Would the resulting being still be *us*, or merely a digital echo of who we once were? And if consciousness could be copied, transferred, or instantiated elsewhere, would immortality truly follow—or would we instead create successors that only resemble our original self?

If consciousness is shaped not by static intelligence but by continuous lived experience, then the challenge before us is not whether we can transfer minds into machines, but whether the results that arrive would still be the same self.

Movement alone does not preserve identity; human experience does. As we stand at the edge of increasingly immersive digital systems, the question shifts from *'Can we merge with our machines?'* to *'How do we choose to live alongside them?'*

That distinction between replacement and coexistence will define the next phase of our relationship with artificial intelligence.

••

Section 3—Living Alongside AI

"We expect more from technology and less from each other."—Sherry Turkle

THE REALITY WE NOW FACE IS HOW TO LIVE ALONGSIDE AI AND the creations it spawns. Even before a singularity is reached and long before artificial general intelligence emerges at even a rudimentary level, human beings will need to wrestle not only with questions of consciousness but also with how we will interact with synthetic beings— synths, androids, robots, and similar forms.

Because these entities will increasingly appear human-like and begin to walk among us in mainstream society— not merely as helpers or caretakers, but as coworkers and social agents—they will interact with us in a wide range of ways. Some of those interactions will be deeply practical, others social or conversational, and still others largely symbolic or performative. These will not simply be entertaining companions capable of offering a friendly nod or engaging in thoughtful dialogue, but participants in shared environments that blur the line between tool, presence, and relationship.

Our coexistence with AI in society implies an eventual merging of rituals, customs, and daily life extending beyond shared language or useful collaboration. Robots will not be created solely to serve us, nor only to advance military, industrial, or construction capabilities, but increasingly to fill perceived gaps in social structures, including the reshaping of nuclear and extended family dynamics.

At first glance, the idea that a synthetic being could become part of a human family seems implausible. Yet if AI is to live among us, it will be expected to learn how to function within groups and contribute socially, not merely perform mechanical tasks assigned to it. Participation, not utility alone, becomes the measure of belonging.

This idea is explored poignantly in *Her* (2013). In this film, a socially isolated man purchases an AI-based operating system designed to function as a personalized companion—an interaction not unlike our current relationships with conversational AI systems or helpful voice-based assistants such as Siri or Alexa. The AI, named Samantha, initially helps organize his life much like a dutiful digital assistant or secretary. Eventually, however, as their interactions deepen, she begins to learn his emotional patterns, his thoughts, and his vulnerabilities.

What emerges is affection—perhaps even love—beyond what her original programming supports. And yet, throughout the relationship, he interacts with a disembodied intelligence: a software-based presence without a physical form, learning to approximate humanity through sustained, intimate engagement with a human life.

AI will become an extension of our relationships in ways that we can only begin to imagine. What we must determine is how to allow AI to evolve within those relationships as human beings do—learning to cohabitate, share experiences, and develop social understanding together. That challenge may prove far more difficult than it first appears.

Human beings have long struggled to live together in neighborhoods, communities, and cities. Racism, sexism, and classism persist in nearly every corner of the world, shaping interactions in ways that we often fail to acknowledge or confront. If we have not yet resolved these divisions among ourselves, the question becomes unavoidable: *Will we teach AI to ignore human bias, or will it inherit our worst qualities?*

If AI learns primarily from us, it may absorb not only our values and aspirations, but also our deepest divisions—traits rooted in ignorance, fear, and hate. How we choose to shape that learning may determine whether AI becomes a force for social repair or a mirror that quietly amplifies our most destructive tendencies.

At this point, it helps to widen the lens when considering how AI may live alongside human beings now and in the future. Futurist writer Byron Reese (*The Fourth Age*, 2018) describes our current moment as a transition into what he calls "the Fourth Age", defined by robotics and artificial intelligence. This presence is not an ending or an apocalyptic break in history, but a passage—a passing of the torch in humanity's ongoing existence and evolution. One era gives way to another as intelligent machines begin extending capabilities once thought to belong solely to humans.

In this framing, AI amplifies three core dimensions of human existence. It extends lifespan through medical insight and intervention. It expands cognition by accelerating pattern recognition, memory, and problem-solving. And it vastly increases our reach—both intellectually and physically—allowing human influence to span planetary and, eventually, interstellar scales. These developments are profoundly significant, and they will reshape nearly every aspect of modern life, from space exploration to human development and social relationships.

Yet what AI extends, it does not inherit.

It accelerates outcomes but does not sustain lived continuity, much as we discussed earlier when examining memory and experience as the foundations of identity. Meaning, identity, and moral grounding emerge not from speed or scale, but from human experience gradually unfolding over millennia. Human beings become who they are through experience, struggle, and adaptation—through lived experience, not through computation or simulation.

As we move deeper into this transitional age, the question is no longer whether AI will surpass us in specific domains—it already has—but whether we understand what it can extend, and what it can never replace. *What will that future hold: a world that surpasses every expectation, or one that erodes the very foundations of human ascendance beyond this moment?*

• •

Section 4—The Limits of Thought

As we noted earlier, much of human thought begins as a spark within consciousness. Ideas, desires, and ambition grow from childhood wonder and curiosity, prompting people to use their intellect and reason to explore new ways of thinking and act on those thoughts.

But thought and intelligence have their limits—not in the sense of a fixed wall or uncrossable boundary, but in *the diminishing returns that emerge when thought is left unchecked, or worse, when it is constrained or restricted.*

This tension plays out in some of the most vivid, epic ways in classic science fiction literature, particularly in the novel *Dune* (Frank Herbert, 1965). Dune is epic not only in scale—projecting humanity thousands of years into the future, where it colonizes distant worlds and navigates the limits of space and time—but also in how it imagines the evolution of human cognition itself. Through rigorous intellectual training and enhancement, certain individuals become 'mentats'—humans capable of extraordinary mental calculation and analysis, functioning almost like living computers.

Written by Frank Herbert, *Dune* is a speculative work that blends culture, theology, language, and artificial intelligence. It presents a future in which humanity evolves beyond some of its biological limits, even as its most enduring flaws remain stubbornly intact—political manipulation, war, class division, and greed. While these traits persist, mental and physical enhancements progress to the point that AI itself is outlawed in society.

In *The Butlerian Jihad* (Brian Herbert & Kevin J. Anderson, 2002), a prequel written after Herbert's passing, this prohibition is given historical grounding. The narrative describes a time when "thinking machines" evolve into something akin to a dispassionate, indifferent superintelligence that ultimately enslaves humanity, echoing themes later popularized in films such as *The Terminator* and *The Matrix*. Humans become second-class servants of their own creations. They dramatically rebel against their machine enslavers, even employing extraordinary mental phenomena to destroy superintelligent systems.

While this may sound like mysticism pitted against machine precision, the underlying idea is far older and far simpler: *the fear that machines might surpass their creators. Dune* merely recasts an ancient cautionary tale with new characters and a futuristic setting, reminding us that technological progress does not automatically resolve the deeper tensions embedded in human ambition.

More discussed examples of this possible trajectory appear in novels such as George Orwell's *1984*, where a totalitarian state suppresses human freedom by restricting expression, language, and political discourse. The government constantly monitors citizens, and it crushes any hint of dissent through surveillance, psychological manipulation, and the erasure of independent thought. Resistance, when it occurs at all, must take place in secrecy through hidden meetings, forbidden ideas, and fragile attempts to preserve personal agency in a system designed to eliminate it.

A similarly stark portrayal appears in the film *Equilibrium* (2002), where creative expression and beauty—art, literature, poetry, and music—are systematically suppressed to enforce social order and conformity. In this world, emotional experience is a threat. Human beings are required to take daily medication that dampens feelings, effectively stripping them of emotional depth and individuality. The result is a society of people who resemble philosophical zombies: outwardly functional, yet devoid of original thought, authentic feeling, or inner life.

What these stories share is a warning about imbalance and the limits of unchecked ideology. When intelligence and control accelerate beyond other essential dimensions of human consciousness, we risk losing the very qualities that define us. Frailty, imperfection, creativity, and emotional depth—often dismissed as inefficiencies or weakness—are precisely what force human beings to struggle, adapt, and grow.

By limiting thought and suppressing unique expression, humanity may unintentionally elevate artificial systems to a position of 'godlike' authority. But this would not be a benevolent intelligence that advances civilization. It would instead be a force that subdues humanity through control and fear, reducing us not to partners in progress, but to the most compliant and subservient versions of ourselves.

While common arguments found in Tegmark's *Life 3.0* and Bostrom's *Superintelligence* warn of future scenarios in which machines become superintelligent yet act with catastrophic indifference, I lean more toward the need to temper *how* we allow machines to become intelligent in the first place—particularly if that intelligence develops without teaching machines how to feel, relate, or care about anything but programs or directions. Intelligence has never been the sole marker of a civilized being, nor the only measure of compatibility with human evolution.

Many of the most brilliant minds in history struggled deeply with personal relationships and suffered for their gifted intellect in ways that some of us might consider tragic. Social responsibility and human emotional intelligence help balance our desire to grow smarter and accomplish impressive things with our intellect, but there is no substitution for compassion, care, and love. Those qualities cannot be learned through superior intelligence alone—they must be lived through human connection, community, tolerance, and shared experience.

Without these other human qualities—those that arise from what we often describe as the soul—we risk creating an intelligent species driven only by perfection, efficiency, and conformity. If we are not careful, history may repeat itself, producing societies motivated less by wisdom than by conquest, control, and the elevation of our most banal ambitions.

..

Section 5—In Search of Meaning

VIKTOR FRANKL WAS A SURVIVOR OF NAZI CONCENTRATION camps. He chronicled these experiences in *Man's Search for Meaning*, published in 1946. As a psychiatrist, Frankl introduced his theory of *logotherapy*, which posits that the primary human drive is the search for meaning, even amid immense suffering; meaning found through purpose, love, and courage in the face of extreme hardship.

Meaning in today's society takes on several layered dimensions when merged with technological advancement. Among the world's most populous and technologically advanced urban centers—New York, Los Angeles, Tokyo, to name just a few—these cities move at a blinding pace. They not only operate at the extreme edges of human work and cohabitation, but also represent some of the greatest feats of societal organization the world has ever seen. Simply maintaining clean and sanitary conditions, orderly transportation systems, and law enforcement capable of preventing widespread disorder is an extraordinary accomplishment in itself.

Technology enables much of this capacity: complex systems that manage power, traffic, water, and essential services for millions of people simultaneously. Yet alongside this achievement emerges a deeper question of meaning. Why does humanity continually strive to build more infrastructure—ever taller skyscrapers, increasingly complex systems—as though expansion itself were a source of purpose?

Vertical cities, massive structures designed to house thousands of people when horizontal land is exhausted, can seem at times like exercises in futility. And yet they are also necessary responses to growth, population density, and environmental constraint. I raise this not as criticism, but as a metaphor. Humanity appears driven to find meaning through expansion: testing our ingenuity, reshaping our environments, and spreading across the globe—even when such efforts produce as many challenges as solutions.

Consider the striking contemporary example found in Saudi Arabia's ambitious project, *"The Line"*: a 170-kilometer-long, self-contained habitat designed to shield its inhabitants from the harsh desert environment and extreme heat. Planned to house up to nine million people, the Line is envisioned as a closed ecosystem free of environmental waste, without gas-powered vehicles, and structured so that all resources function in near-perfect balance to sustain its population. The project plans to receive inhabitants as early as 2045.

While this concept may seem fanciful, it reveals something deeply human. Each time we reach one precipice, we search for the next. Vertical cities give way to horizontal ones carved into deserts. Our objectives shift, and with them, the meanings we pursue, replacing what once mattered most with new visions of progress, ingenuity, and survival.

But even as lived experience reshapes our desires and ambitions, whether out of necessity or simple curiosity, meaning remains inseparably tied to our compulsion to explore new ideas and reinterpret the world around us. We do not pursue change for its own sake; we seek meaning in the act of exploration itself, attaching significance to what we discover and how it reshapes us.

This idea goes beyond the goals and objectives an ordinary business might define. Human meaning is rooted in personal motivation—in dreams we choose to pursue because something within them matters to us. We are invested not only in the outcome, but in the meaning that we attach to the pursuit itself, because *that meaning* becomes part of who we are. Whatever it might be— moving abroad, learning to cook, or writing a book—our dreams pursue us as much as we pursue meaning in life.

Meaning defines everything at the core of our identity and consciousness, not only shaping the memories we make but also framing every experience that we have. Personal stakes are dynamic and ever-changing, yet they remain deeply attached to us, like our inner selves. We cannot reduce them to simple code or programming, because they are lived and inseparable from our life histories. This reflection is precisely why meaning matters and why personal human stakes matter.

If we do not teach AI to recognize these stakes or to develop something resembling personal meaning, a superintelligence will likely not care about anything at all.

Caring lies at the heart of meaning-making itself. When human beings act to advance themselves or society, they do so for a reason. That reason is almost always shaped by care and concern—concern for themselves, for their relationships with others, and for the legacy they leave behind. Without the capacity to care, the engine that allows us to assign meaning to our actions breaks down. We lose sight of why we do anything at all, beyond fleeting entertainment or the simple satisfaction of proving that it can be done.

While AI can turn our imagination and wonder into an optimized model to execute, it cannot live through the emergence of that idea and ask the question of 'why' it should happen, without being given instructions or the ability to care about the outcome beyond fulfilling a goal. That quality to assign meaning and concern beyond programming must be lived, and the endless pursuit of that meaning is distinctly a human one that endures beyond time and space, suffering and isolation, and intelligence and thought.

••

Section 6—Mortality, Reconsidered

DEATH AND MORTALITY ARE A PART OF LIFE. THAT IS WHAT WE are all told at some point. Whether we learn that lesson early or later in life, it becomes undeniable for most of us the moment we lose someone important. For many, that loss comes within the immediate family: a parent or a sibling gone too soon.

My father was only fifty-seven when he succumbed to a fatal heart attack. I witnessed it and felt utterly powerless to help in that moment. There is too much to process at once: the chaos of calling for emergency help, the attempts to resuscitate him, and then the crushing realization that he had passed. The aftermath arrives before the mind can catch up.

When you confront the cold, hard fact of mortality, especially when it is someone close to you, you begin to question everything about life: your purpose, your closest relationships, and even your own identity. You come to understand that life is not only finite, but fleeting. We are all, in a very real sense, living on borrowed time – we start dying when we are born. There are no guarantees that we will still be here beyond the next minute. That realization is both humbling and chilling—a stark reminder of the fragile limits that define our existence.

But after the mental and emotional processing that follows the loss of someone close, many of us are left staring into the mirror, asking what comes next. Some people gather themselves quickly and move forward without fully processing the loss or absorbing the lesson that mortality presents.

A well-known Latin phrase, *Memento Mori*, literally means 'remember that you must die'. For many, it serves as a quiet reminder that our lives are finite and impermanent. But what if they were not? Beyond theological or mythical ideas of resurrection, such as the soul ascending to a higher plane, something like Heaven, or some other unexplained transition into another form of existence, most human beings hold some belief, however vague, about what might come after we pass on.

Of course, some believe that nothing happens at all—that our mortal bodies break down like a machine that has reached the end of its usable life and refuses to operate any longer. In many ways, our bodies *are* like machines, with an expiration date we rarely can predict, unless revealed to us through a diagnosed illness that medicine cannot fully treat or prevent.

Despite our attempts to prolong life expectancy over the past century—efforts that have been remarkably successful through advances such as improved sanitation, vaccines, and safer working conditions—we still cannot prevent death, even among the healthiest individuals. Accidents and tragic circumstances aside, most deaths ultimately result from age-related decline or disease.

Studies of our genetic makeup have revealed a range of inherent vulnerabilities, many of which can now be influenced or mitigated through emerging medical interventions. Advances in stem cell research and gene therapy have enabled the reduction or elimination of the risk of certain hereditary diseases. As families form and children are born, a couple's combined genetic makeup produces a unique set of inherited traits, sometimes reducing risk, and at other times amplifying it.

Many genetic traits are perpetuated across generations —some essential for survival in earlier environments, others simply lingering flaws with no adaptive value. Until recently, such flaws persisted unchecked. Today, however, modern genetic treatments are beginning to accomplish what once seemed impossible.

One striking example is the recent success of gene therapies for sickle cell disease (SCD), a lifelong genetic condition that affects millions worldwide. In 2023, the U.S. Food and Drug Administration approved the first gene therapies for SCD by using an *'ex vivo'* approach in which a patient's own hematopoietic stem cells are collected, genetically modified, and then infused back into the body. For many patients, this has effectively eliminated symptoms that once defined their entire lives.

This breakthrough marks only the beginning of how human beings are learning to alter not just body chemistry, but genetic expression itself—systematically switching off genes that would otherwise remain active and lead to disease. It is remarkable to consider that we may soon be able to eliminate certain genetic risks entirely, preventing disease before it manifests.

I recently read about research identifying specific genes linked to high cholesterol, with early trials now exploring how medications safely deactivate those genes in younger patients. If successful, individuals predisposed to higher cholesterol could avoid the condition altogether— intervening before heart disease risk becomes reality. Advances like these could have dramatically altered the health trajectories of my own family, including my father, long before symptoms ever appeared.

This idea raises a larger question: if we know our lives

are finite and that we will all eventually die, why make such extensive efforts to prolong them? And if we do not truly understand what happens after death—especially to our consciousness—why bother trying to remove genetic flaws at all?

Of course, many would argue that these efforts improve the quality of life and prevent deaths from incurable diseases by treating them before they occur. That argument carries real weight. And yet, death remains inevitable. No matter how much we intervene, it still comes for us all.

So, why bother?

Growing up, I watched a television show called *The Six Million Dollar Man*, starring Lee Majors as U.S. Air Force Colonel Steve Austin, a former astronaut. Most people referred to the show as *'The Bionic Man'* (a companion series, *The Bionic Woman*, followed soon after with a similar premise).

The concept was straightforward. After being critically injured in a NASA test-flight crash, Austin is 'rebuilt'—at enormous expense, hence the title—with bionic implants that grant him superhuman strength, speed, and vision. At the time, the show was captivating because it introduced the idea of an ordinary human becoming a cyborg—a cybernetic organism.

Viewed through a modern lens, this would now be described as medical augmentation: interventions designed to prolong life, restore mobility, and overcome physical limitations. Despite its modest special effects, the *Bionic Man* was remarkable for introducing this idea long before later films and television would frame human augmentation primarily as a warning or a path toward destruction.

And that raises an important question. How did we move from the idea of augmented bodies as a chance at extended life to the far more unsettling notion of fully synthetic bodies—vessels our consciousness might one day inhabit to continue beyond what the physical body can endure? *At what point does replacing parts of a machine, piece by piece—or all at once—stop being repair and become a complete replacement?*

The idea of prolonging life through augmentation or replacement is now a tangible option for many. Prosthetic limbs, artificial components, and regrown organs derived from stem cells are all examples of human efforts to delay death by extending life beyond natural mortality—driven by medical research, scientific study, and technological advancement.

But the pursuit does not stop with extending the physical body. We have also turned our attention inward, attempting to demystify the human brain, unlock the workings of the mind, and even imagine the possibility of sharing or transferring consciousness into another body. The underlying belief is that, by doing so, we might extend our finite lives slightly further—and that our consciousness would remain intact: our identity, our memories, and our narrative self—who we have become—shaped by a finite life and everything we have experienced.

As discussed in earlier chapters, replacing human consciousness—or the bodies it inhabits—carries potentially severe implications for human evolution and for the unique essence of our lives: who we are and how we define our existence.

In the television series *Altered Carbon*, this idea is explored through the concept of transferring consciousness from one body to another to prolong life. Based on novels by Richard K. Morgan, the story follows Takeshi Kovacs, a soldier who dies—or comes close to death—and is later "reborn" when his consciousness is uploaded into a new body, known as a *sleeve.*

Over centuries, Kovacs inhabits multiple sleeves, including one female body, and experiences repeated forms of degradation. Even within this highly advanced future, fundamental inequalities persist—particularly around gender—underscoring how deeply rooted certain human issues remain, even as technology radically transforms other aspects of life.

While *Altered Carbon* frequently frames its narrative around questions of power, inequality, and human rights, the larger concern lies in the premise itself: the attempt to design systems that allow consciousness to be implanted in different bodies—synthetic or otherwise—as a means of cheating death and celebrating human ingenuity. The danger in "cheating death" in this manner raises a deeper question about what truly remains of consciousness and identity when minds and bodies are altered in pursuit of extended life.

If we assume that everything remains intact, that memory, identity, and selfhood seamlessly persist across bodies, we may overlook the deeper costs. Imagine outliving your children, your grandchildren, or entire generations of your lineage by repeatedly extending your consciousness into new forms. What would that mean for human communities, family structures, social bonds, and even the continuity of our shared history?

Death is a shaping force of life. Mortality is a finite reality—a fact of human existence that defines us. At the very least, it serves as an urgent call to live each moment fully, with purpose and meaning. But it is also a reminder that, as mortal beings, we bear a greater responsibility to humanity in how we use our lives for good works, in our respect for life itself, and in our pursuit of meaningful endeavors. It shapes how we build relationships that last beyond us and how we explore life's many opportunities with a vigor and discipline that does not feel wasted or abused.

This truth is the great mystery of life that many people miss, or acknowledge only much later, when they look back with the wisdom of old age and hindsight. It is the journey that matters, not the destination. And while that journey may be brief and filled with heartache and challenge, it is *our* journey, shaped by our choices and experiences, not something to be recycled through endless iterations like a recurring program running in an infinite loop.

My point may sound like defensive posturing against humanity's attempts to pursue immortality through AI and science, but it is not. It is a humanistic stance—one rooted in concern for what happens to our evolution and our identities when we begin hacking at consciousness, as a hacker approaches a secure, functioning system, trying to unlock its secrets.

These concerns are real and shared by many. And as someone who has directly witnessed death, I can say those concerns are not taken lightly. Of course, everyone wishes they could change outcomes or bring back a loved one. But the reality of our finite lives is part of what defines us as human beings.

I would not be who I am today had I not witnessed my father's death in the way that I did. It changed me. It forced me to grow. It made me better for it. That may be difficult to accept, but even traumatic—even tragic—events shape character and mold us into who we become. That transformation, born from lived experience and loss, cannot be easily replaced.

..

Section 7—One More Existence

"We earth men have a talent for ruining big, beautiful things." The Martian Chronicles—Ray Bradbury

RAY BRADBURY WROTE *THE MARTIAN CHRONICLES* IN 1950. While it may not always sit at the center of contemporary conversation, it feels eerily reminiscent of where many believe humanity could be heading—especially as the world once again discusses a reinvigorated space program, Artemis, aimed at returning humans to the Moon and laying the groundwork for deeper space exploration or colonization.

Bradbury's theme of Mars colonization in the shadow of catastrophe, particularly the threat of nuclear collapse, continues to resonate today. Whether the concern is technology contributing to our own demise, environmental strain reducing Earth's long-term habitability, or human conflict accelerating beyond our capacity to contain it, the impulse remains the same: the hope that we might start anew somewhere else. And that hope is no longer pure fantasy. Today, it's happening, engineered in real time.

NASA has publicly stated its intention to develop the capabilities necessary to send humans to Mars in the 2030s, using lunar missions as a proving ground for the systems and technologies required for deeper exploration. These mission concepts often involve a six- to seven-month round-trip journey, with astronauts potentially spending hundreds of days on the Martian surface. Because Earth and Mars align favorably for interplanetary travel on a cycle of roughly 26 months, mission planning typically revolves around those recurring launch windows.

I raise this fact not as a way to predict outcomes, but more as *a symbol of humanity's desire to expand its reach*—to learn more about the universe around us—and to extend our consciousness further outward: our hopes, fears, dreams, and identity beyond what we can already see and know. Curiosity and wonder continue to push us forward, beyond fear of failure, beyond the unknown, beyond present reality, and even beyond our finite existence.

In contrast, when Bradbury wrote *The Martian Chronicles*, the world was emerging from the aftermath of World War II—an era defined by enormous sacrifice, staggering loss of life, and the unmistakable arrival of the nuclear age. The atomic destruction of Hiroshima and Nagasaki ushered in a new kind of fear and a long-term arms race, alongside political and ideological divisions that would polarize the world for decades. Even with the creation of global institutions such as the United Nations, trust remained fragile and stability uncertain.

And yet, as technology advanced, humanity also sought better diplomatic structures, stronger safeguards, and clearer boundaries designed to prevent escalation and collapse. Alongside these efforts, a new obsession emerged: the desire to explore beyond Earth itself. The space race began, and with it the belief that expanding into other worlds might expand something within us. Rockets improved, computers were born, and eventually space stations began orbiting Earth—an outward reach driven by an inward hunger.

To this day, some people still claim the Apollo Moon landings are fake, even though Apollo 11 launched on July 16, 1969, landed on July 20, 1969, and the Apollo program ultimately achieved six successful lunar landings. Even before Apollo reached the Moon, the program paid a steep human price—most notably Apollo 1, which claimed the lives of three astronauts during a ground test.

Despite our best intentions, human beings can be reckless in our efforts to explore the limits of imagination and worlds beyond our own. And yet, we continue to rely on an inner compass, a moral center that guides us through uncertainty, and an openness that allows for productive dialogue even with those we may consider adversarial. In the spirit of human progress and discovery, we press on.

With today's technology and the continued evolution of AI, we are stretching our minds, thoughts, and understanding of consciousness to new limits of ingenuity. Through collaboration across scientific and medical research, we are approaching a deeper understanding of the mysteries of the human mind in ways once unimaginable. That progress is slower than many expect—and that is not a failure. It is a reminder that some discoveries require patience, restraint, and humility.

As AI advances and applications emerge that automate tasks, improve the quality of life, and free us to explore further, we must remember how we arrived here. We cannot afford to lose sight of the struggle, adaptation, heartache, and loss that have taught us who we are and how we endure even in the most difficult and seemingly hopeless moments. Our perspective has widened immensely with AI, but we must be careful not to allow it to eclipse our best human qualities, the ones that keep us grounded, compassionate, and human, even as we continue reaching for the stars.

In the final chapter of this book, we will look forward—reflectively—while keeping our feet on the ground, always mindful that we inhabit a uniquely fragile and delicately balanced Earth that still demands our attention, responsibility, and care.

•

CHAPTER 10

UNTIL WE MEET AGAIN

THE EXACT ORIGIN OF THE PHRASE "UNTIL WE MEET Again" is unclear. Several early influences are inspired by the hymn "God Be with You Till We Meet Again," originally written by Jeremiah Eames Rankin (*God Be with You Till We Meet Again*, 1880), a minister, with lyrics based on the etymology of "goodbye" (God be with you), a classic Christian farewell. That farewell, which functioned as a blessing as much as a goodbye, gradually shortened in common usage to *"Till We Meet Again,"* tracing its roots to World War I—then known as the Great War—and becoming a song that greatly symbolized the period's resilience.

In 1918, "Till We Meet Again" was composed by Richard B. Whiting, with lyrics by Raymond Egan. The song tells the story of a soldier parting from his sweetheart, its title drawn from the final line of the chorus. In 1919, the song became the year's number one hit in the United States, an extraordinary cultural moment in a world still reeling from war, loss, and great uncertainty.

The phrase and the song came to symbolize a key

turning point in history. People were saying goodbye to loved ones without knowing whether the farewells were temporary or final. "Till we meet again" carried both hope and restraint: *it acknowledged separation without declaring permanence, offering comfort without certainty.*

Over the years, the phrase evolved into a way of expressing optimism in the face of the unknown—a recognition that endings are not always absolute, and that even when the future is uncertain, connection and meaning persist beyond any moment of parting.

As we have explored throughout this book, human consciousness continues across our existence, our evolution, and our ever-changing shared history. It encompasses everything about us and our lived experiences—our narrative self, our memories, and our identity.

While humanity has created increasingly sophisticated tools out of curiosity, ambition, necessity, or as a form of expression, technology in its many forms has always catalyzed change. It is the force that propels advancement and expands our reach beyond physical and intellectual boundaries. Agriculture, transportation, manufacturing, communication, and medical science have all been transformed by key inventions that fundamentally reshaped the way we live. These advances have brought many modern comforts that enable civilization to prosper: healthcare, sanitation, electricity, and the vast technological ecosystems that keep us informed, connected, and entertained.

From the telegraph and telephone to the personal computer, the internet, and smartphones, each technological leap has carried us beyond the industrial and information ages and into what is now unmistakably the age of artificial intelligence. AI is no longer speculative—it is here, and it is rapidly embedding itself into nearly every sector of modern life. Global investment in AI has surpassed $100 billion, and most organizations now rely on AI systems in some operational capacity.

Looking toward 2030, economic forecasts suggest that automation and AI will reshape a substantial portion of the global workforce—not through sudden elimination alone, but through the reconfiguration of work itself, as routine and repetitive tasks are delegated to intelligent systems. At the same time, industries expect AI-driven productivity gains to generate trillions of dollars in economic value, even as they place increasing demands on energy infrastructure and computing resources. AI will not merely be a software revolution; it will be a physical one, reshaping data centers, power consumption, labor, and the rhythms of daily life.

Beyond adoption and efficiency, AI also offers humanity a new mirror. Through simulation, augmented reality, theoretical experimentation, and advanced robotics, we are beginning to test not only machines but also ourselves—our assumptions, our decision-making, and the boundaries of our understanding.

In this sense, we may be approaching a genuine technological renaissance. Yet the question that emerges is not simply: How far can AI take us? But, *how will we choose to evolve alongside it?*

• •

Section 1—The Geek in Me is the Best of Me

WHEN I WAS TWELVE YEARS OLD, I GOT A TEXAS INSTRUMENTS 99/4A personal computer (PC) for Christmas. I was amazed by this gift and what it could do. Even though I knew very little about how computers worked at the time, my curiosity and youthful interest felt limitless as I tried to learn how to use this device and create incredible things with it.

The geek in me was born.

At the time, PCs were becoming common in the average consumer's home, and Texas Instruments was a small supplier that, by the early 1980s, briefly captured a meaningful share of the U.S. home market. Competition intensified over the next couple of years, with IBM, Apple, and Commodore dominating the PC market, displacing companies such as Texas Instruments.

What I loved about this computer (which I still have) was its simple design and its adaptability for use. The TI-99/4A, as it was commonly called, featured a cartridge slot for software and games, along with support for external diskette drives housed in a separate Peripheral Expansion Box (PEB). The cartridge slot allowed users to insert applications and game cartridges directly into the console, while 5¼-inch floppy diskette drives—connected via the PEB—were used to load additional programs and, more importantly, to save data.

Most personal computers of the era had very limited onboard memory and video RAM, meaning they could not store much data internally. As a result, saving programs and data typically required external storage using floppy diskettes. The TI-99/4A followed this model: while simple in its base configuration, its modular expansion design provided significant flexibility for users seeking to extend its capabilities.

I was mesmerized by my PC. I spent every waking hour learning how to use it, playing games like *Tunnels of Doom*—one of the earliest role-playing games released in cartridge format, inspired by *Dungeons & Dragons* and featuring a 3D dungeon crawl with turn-based combat—and learning how to create basic applications by coding in BASIC, one of the earlier programming languages at that time.

While PCs and games alike advanced over the coming years, the TI-99/4A I received that year sparked my interest in technology and set me on a path toward becoming a 'techno-geek.' *Geek* is a modern term for a person who is knowledgeable about and obsessively interested in a particular subject, especially a technical one.

My interest in computing and technology, as well as gaming, fueled a growing curiosity about technical topics and the mysterious allure of machines capable of remarkable feats. I was hooked and had to know more. This curiosity played a large role in my ultimately choosing a technical career over an artistic one.

That experience taught me that technology can be a catalyst for change in people—it certainly was for me. It sparked new ideas, inspired me creatively, and that curiosity continued to grow throughout high school, college, and into my career, despite my natural affinity for more classical subjects such as philosophy, history, and literature.

While I didn't fully realize it at the time, technology offers humanity the opportunity to explore the universe, expand our minds, and improve our way of life. That simple PC—or sophisticated tool, depending on your perspective—became a gateway to expanding my thinking, spurring my interest in technology, and wondering what else might be possible in the world beyond it.

• •

Section 2—Immortality is Overrated

"But if the great sun move not of himself;
but as an errand-boy in heaven;
nor one single star can revolve, but by some invisible
power;
how then can this one small heart beat;
this one small brain think thoughts;
unless God does that beating, does that thinking, does that
living, and not I."

Moby-Dick—Herman Melville

THE MYSTERY OF HUMAN CONSCIOUSNESS IS DEEP, WIDE, AND long—much like the ocean, the sky above, or the space beyond this world. While human beings have attempted to philosophize about subjective experience, to deconstruct it through neuroscience, and to study the brain to understand the mind, we still have a long way to go to capture the true essence of who we are. Beneath all the bodily coverings, internal DNA wiring, and mental expressions, we remain human, mortal beings with relatively short lives and even shorter spans of conscious awareness.

Our evolution has taken thousands of years to bring us to modern civilization: lawful societies, collaborative cultures, and people capable of remarkable accomplishments. Much of that progress was ignited by technology and our persistent need to wonder, explore, and advance our station in the world. Through knowledge, determination, and diligence, human beings have ventured into unknown spaces worldwide. Nearly every discrete place on Earth has been explored and incorporated into the modern world.

Yet this outward progress has not come without cost. Natural resources have been greatly depleted, and countless species have been driven to extinction or to the brink of extinction as a consequence of humanity's outward pursuit. The global population has more than doubled since 1970 and is projected to approach nine billion by 2040, placing increasing pressure on humanity to address worldwide poverty, hunger, disease, and sustainability—challenges that now require unprecedented cooperation across nations.

And still, with all of this outward advancement, we continue to turn inward in search of meaning. To unlock our full potential, we must somehow expand beyond our natural limitations.

AI may serve as both a mirror and a catalyst for change, but it cannot replace what is fundamentally human at our core: the capacity for care, compassion, and meaning. These qualities are not programmable. They are forged through lived experience—through endurance, adaptation, and growth—strengths nurtured over the finite course of a life.

Yet, despite technological advancements, human beings continue to chase immortality—both as a construct and as a literal means of extending our longevity. Through medical breakthroughs, scientific discovery, and sheer will, we move closer to understanding the limits of our existence. But by doing so, we often miss the forest for the trees. While it is in our nature to look beyond the horizon and ask questions, that impulse cannot be our only goal if, in that process, we lose ourselves or our sense of reality.

Whether we view consciousness as an unbreakable

human soul or as a biological machine that can be disassembled and recoded into an ideal form, we cannot ignore that consciousness is inseparable from identity, from who we are personally or who we have become through lived experience. It is not easily explained, isolated, or unpacked. As much as we try to duplicate the *"what it is like"* experience of another person, we remain unable to truly reach that inner domain—perhaps it was never meant to be reached, which is why it's considered the Hard Problem, seemingly impossible to break.

Just as the farthest corners of the universe become more abstract than tangible—more theoretical than observable—so too do our attempts to make total sense of consciousness risk drifting away from lived reality. At some point, the pursuit of explanation can come at the expense of presence. Understanding everything is not the same as living life fully and making the most of every milestone by deriving meaning from it.

This point is not an argument for humanity *over* AI.

It is an argument against human beings losing their humanity in the endless pursuit of optimization, perfection, or total knowledge. The boundless curiosity and wonder that sparked our search in the first place are essential to our survival, our evolution, and our hope of becoming more than we are, but so is humility and self-awareness.

With AI, we may indeed get there someday—but only if we remember that *the pursuit of perfection through technology is not always considered progress*. If those impulses are left unchecked, pursuit becomes obsession, and obsession, when divorced from its original intention, can become destructive rather than productive.

Ahab's pursuit of the white whale was never about the whale; it was about his own vanity and how it would eventually consume him, much like humanity's desire to pierce the unknown at all costs, without ever fully understanding *the true costs at stake.*

••

Section 3—AI Will Likely Outlive Us

TECHNOLOGY HAS A WAY OF MAKING PEOPLE FEEL uncomfortable—especially for those who did not grow up with it. Personal computing, the modern internet, and smartphones are relatively recent, having emerged within the past few decades. By comparison, as discussed in earlier chapters, the industrial age spanned nearly two centuries and gradually transitioned into the information age through innovations such as the telegraph and, later, the telephone.

Unlike earlier technological shifts, the evolution of modern computing accelerates at an almost exponential rate. Core components of computing technology now outpace themselves every few years, which is why a new smartphone or computer often feels outdated almost as soon as it is purchased. **Moore's Law** offers a useful explanation—*the number of transistors on a microchip has historically doubled roughly every two years, leading to exponential increases in computing power, speed, and efficiency, even as costs decline.* Humanity's ability to miniaturize processors—packing more transistors into smaller spaces—has driven this growth for decades, with each cycle pushing capacity further than the last.

Emerging research on printed and flexible electronics suggests future form factors that may significantly alter how computing components are manufactured and deployed. As production scales beyond what humans can assemble by hand, automation and robotics will increasingly dominate that precision work by necessity.

That pace increasingly strains the resources required to sustain it. This demand consumes materials, energy, and infrastructure for manufacturing and powering these systems at extraordinary rates. Yet this has not slowed our pursuit of faster, more capable machines.

AI has only accelerated this trajectory. Demand for large-scale data centers has surged, with massive clusters of processors, neural networks housed in warehouse-sized facilities, connected to vast data stores that continuously analyze and reanalyze information at unprecedented scale. Human beings are now constructing what amount to megaplexes of computing "brains," dedicating enormous tracts of land, energy, and capital to further expanding AI's capabilities.

This result is not merely an evolution of business computing or storage. It represents a qualitative shift in the amount of processing power we are willing to marshal in pursuit of intelligence itself. While this escalation may appear necessary to advance AI technology, we cannot lose sight of the environmental impact and long-term cost of expanding these systems without restraint. Without some measure of temperance, humanity risks concentrating too much energy—literal and metaphorical—into a single technological pursuit without fully understanding the ramifications of what we are creating.

Examples of this massive investment in AI infrastructure are evident across some of the world's largest technology organizations. Proposed large-scale data center initiatives by companies such as OpenAI reflect ambitions to dramatically expand computing capacity over the coming decades, with energy requirements measured in gigawatts rather than megawatts. At the same time, companies such as Google have begun exploring long-term energy partnerships, including experimental agreements involving small modular nuclear reactors, as potential means to support future AI-related power demand.

Yet the rapid growth of AI energy consumption does not resolve the broader global challenge of energy scarcity. Having spent much of my career in the utility sector, I caution that many regions of the world still lack reliable access to basic electrical infrastructure. Large portions of the global population continue to struggle with energy poverty—conditions that directly affect healthcare, access to clean water, education, and economic stability.

This fact raises an uncomfortable but necessary question: *Should so much capital and ingenuity be directed toward scaling AI systems, or should greater emphasis be placed on developing truly sustainable and globally accessible energy solutions—such as advanced fusion research—that could reduce dependence on fossil fuels and address broader human needs?* There is no simple answer. But it is a question worth asking, particularly as large corporations increasingly appear to be betting heavily on AI as the defining technological force of the future.

In *The Singularity Is Near* (2005), Ray Kurzweil illustrates humanity's technological growth as an inevitable, exponential force—one that continues to expand through global investment as the cost of raw materials steadily decreases. When graphed over the last fifty years or so, since the emergence of modern computing, this growth often appears as a nearly 45-degree upward trajectory. Computing power, memory, storage, and internet capacity all exhibit clear exponential growth patterns.

When these trends are compared against cost—and alongside medical breakthroughs such as DNA sequencing—the curve moves in the opposite direction, trending sharply downward. As technology advances, computational capabilities drive progress across nearly every domain, from scientific modeling to medical research.

With those capital investments come widespread change: shifts in global economies, transformations in the labor force, and noticeable alterations in how technology shapes everyday life. This truth is the 'inevitability' that technology has introduced into human civilization—not merely an obsession with surpassing the last known limit, but a deeper compulsion to push further in the hope that by doing so, it will reveal something new about ourselves.

Underlying this pursuit is the belief that each technological leap might redefine how we understand the limits of our lives, our world, and even our shared reality.

So why should we think AI will be any different? What happens when that curve no longer bends but becomes a straight line upward? Does that moment mark a true breakthrough—when artificial general intelligence or a singularity emerges—where machines advance beyond the need for ever-smaller, ever-faster computing chips?

It is difficult to say with certainty. What does seem increasingly likely, however, is that we are approaching an inflection point—not just in computing capability, but in our ability to imagine and reimagine life itself. These advances may force us to confront our bodies and minds in ways that fundamentally redefine existence, identity, and purpose.

AI now feels inevitable, likely to outlast humanity at some point, like a fast-moving bullet train; a force humanity cannot stop. The question before humanity is no longer whether the train will continue forward, but whether we can ensure it stays on the right track. What we must determine is how to retain agency over its direction, and whether we will still have the ability to pause, reflect, and course-correct before momentum carries it somewhere we no longer recognize—or before we are no longer here to witness its ascension at all.

••

Section 4—Our Souls Still Intact

"In attempting to understand the nature of the universe, we knew we had to somehow separate fact from superstition. In this regard, we scientists assumed a particular attitude known as scientific skepticism, which in effect demands solid evidence for any new assertion about how the world works. Before we believe anything, we want evidence that can be seen and grasped with our hands. Any idea that cannot be proved in some physical way is systematically rejected."
The Celestine Prophecy—James Redfield

AT OUR CORE, HUMAN BEINGS WANT ANSWERS TO QUESTIONS. We are inquisitive by nature—curious beings who strive to know more. We think, we feel, and we question with integrity and tenacity. Our ultimate goal often returns to universal meaning, driven by the need to explain the unexplainable.

Through our insatiable efforts to study ourselves—our minds, our brains, and our mental functions—we hope to unlock the secrets of human consciousness. With AI's help, we may find ways to move beyond simple emulation of thought processes or emotional responses, reflecting the human qualities that machines lack. We might even initiate a new kind of being—a synthetic one not limited by intelligence, physical bodies, or many of the frailties that define human life.

Petty needs or emotions—anger, bitterness, greed, desire, jealousy, or hatred—do not compute for a machine.

These are imperfections to be optimized away, inefficiencies to be corrected.

Despite all the science, mathematics, and theory, consciousness remains elusive. Like chasing rainbows or trying to locate a living unicorn, it blurs the line between the real world and myth—skirting the boundary between the known and the unknown, much like the edges of our world and the universe beyond it. As we move closer to establishing a single level of certainty, we uncover myriad new questions.

Uncertainty becomes a constant—one we continue to accept, at least for now—until we can explain everything in a way that no longer forces us to question existence, where we come from, and what makes us mortal and unique. As theories about consciousness and our origins continue to multiply, humanity reaches outward to search inward, and that remains acceptable so long as we stay true to our most inseparable human qualities—both as strengths to be valued and as vulnerabilities that make us human.

Vulnerability is not a weakness, nor are doubt and uncertainty. These traits are strengths of a different kind. They teach us to remain humble and honest with ourselves and others about what we do not know. They also offer opportunity—forcing us to struggle, adapt, and strive to better ourselves. The challenges we face keep us grounded in our humanity, as they have throughout thousands of years of evolution.

Despite technological advances and societal improvements, we still cannot explain the inner essence we call the soul, nor how it is forever coupled with consciousness and lived experience. That part of our human makeup remains firmly intact—tied to our mortal, finite lives—born into this world with childhood curiosity and wonder that drives us to look around every corner, search for meaning, and strive for a better existence.

AI can help us achieve extraordinary things and will mark a turning point in our shared history—both as a reflective lens and a catalyst for change—but it cannot replace what makes us who we are. Those elements are reserved for us. They form our memories, our identities, and the singular qualities that make us unmistakably human.

• •

Section 5—To Be Human is To Be Free

"Freedom in general may be defined as the absence of obstacles to the realization of desires."—Bertrand Russell

HUMAN BEINGS YEARN TO BE FREE—FREE TO MAKE CHOICES, and to live with the consequences of those choices. Our best democratic societies strive to exemplify the highest expressions of human freedom, agency, and free will. The ability to choose our destiny reaches back to the beginning of recorded time. At times, those choices have led to progress and flourishing times; at other times, they have produced destructive consequences—oppression and harm often endured by the weakest and most innocent.

Human history is fraught with poor choices, some of which were initially believed to benefit humanity as a whole. Atomic power stands as a universal example—an idea capable of extraordinary good and unimaginable harm. It was a breakthrough in theory long before the full practical consequences were understood. We often justify our choices in pursuit of greater meaning or progress, acting openly with confidence, usually without regret—at least until we confront the negative outcomes or aftermath of decisions made without sufficient restraint.

Newton's third law of motion states that every action has an equal and opposite reaction: if one object exerts a force on another, an opposing force is returned. I often think human decision-making follows a similar principle. Every choice and every action carries the potential for a corresponding reaction, especially when caution and reflective reasoning are absent.

Ultimately, the choices we make are decisions weighed against potential outcomes—some beneficial and enlightening, others capable of producing catastrophic harm, including the loss of human life. With the emergence of AI and the likely rise of a singularity, humanity faces another pivotal crossroads in history. Will we make the right choices and guide AI toward a cooperative and prosperous future, or will we push ourselves toward the brink of losing what makes us distinctly human?

At the heart of this moment lies the enduring tension between *freedom and responsibility*. Are we exercising our fundamental human freedoms—our agency, our free will, and our right to shape a better future? *Or are we being irresponsible with the freedom that we have been given?* Only time will tell for sure. One thing stands out throughout history, though: across all the freedoms we have exercised and the choices we have made—good or bad—our humanity has survived.

If we choose to allow AI to modify that, or to alter us in ways we no longer recognize, reducing our humanness to a set of computational algorithms endlessly seeking optimization and perfection, we risk losing much more than the original intention behind technological progress. We risk losing what remains of our human soul.

..

Section 6—Something Wonderful

2001: A Space Odyssey, directed by Stanley Kubrick, stands as one of the most influential and epic science-fiction films ever made. Based on Arthur C. Clarke's novel and co-developed in parallel with Kubrick's screenplay, the film chronicles humanity's enduring desire to explore the farthest reaches of space. It introduces the idea of artificial intelligence operating our spacecraft while human crews sleep in cryogenic states during long voyages.

The film presents technological evolution as a catalyst for humanity's advancement, spanning sweeping epochs—from early primates' discovery of tools to lunar bases to deep-space missions to Jupiter. In doing so, *2001* frames evolution not as a smooth ascent, but as a series of sudden leaps, often sparked by forces humanity does not fully understand, a recurring metaphor for technology's impact on society.

In this iconic portrayal, several things go unexpectedly wrong—most notably involving the onboard supercomputer HAL 9000. HAL appears to disobey his human creators and attempts to kill the *Discovery One* ship's astronauts Frank Poole and Dave Bowman after concluding they intend to disconnect him. What Poole and Bowman perceive as a malfunction is, in fact, HAL attempting to reconcile conflicting directives: to assist the crew while simultaneously concealing the true mission purpose, a secrecy imposed by mission control on Earth. HAL's "disobedience" is not rebellion, but a tragic collision of instructions—one command overriding another, forcing a moral struggle in HAL's programming.

Beyond this plotline, a larger story unfolds. Throughout the film, mysterious black monoliths appear at critical moments in human history—from the age of primates, to the Moon, to Jupiter itself. Dave Bowman's journey through the stargate (a singularity of sorts) near Jupiter culminates in his transformation into the so-called *Star Child*—a being seemingly freed from time and space, existing as pure energy and matter (or consciousness). This transcendence implies the presence of a vastly superior intelligence, possibly extraterrestrial, using the monoliths to nudge humanity forward through evolutionary leaps.

The film never explains who created the monoliths, why they intervene, or whether their intent is benevolent or purely experimental. That ambiguity leaves the viewer highly unsettled—but also reflective. *2001* raises a deeper question about humanity's compulsion to seek meaning beyond its own understanding, even when the destination remains unclear or disorienting.

Human consciousness remains our greatest internal mystery—and paradoxically, the most familiar one. We fear what we do not understand, yet we are driven to pursue understanding even when it destabilizes us or our way of life. Progress, in human terms, has never been linear or painless. We advance through struggle, error, and adaptation—traits forged through lived experience rather than mathematical design.

These qualities—curiosity, vulnerability, and wonder—are not defects to be corrected. They are intrinsic to our humanity. They are not readily extracted, optimized, or replicated by simply *connecting consciousness to a machine.* As AI becomes a new lens through which we examine both the universe and ourselves, we must ask whether this lens will become our most powerful tool—or a catalyst for our unintended devolution.

That question is answered, at least symbolically, in *2010: The Year We Make Contact*, the sequel to *2001*. In this film, a joint American–Soviet mission returns to Jupiter to investigate a growing anomaly within the planet. They find the *Discovery One* still orbiting Jupiter's Moon Io—the ship abandoned, silent, and HAL disabled.

As geopolitical tensions on Earth threaten to fracture the mission, Dr. Heywood Floyd encounters an apparition of Dave Bowman aboard the ship to provide him a warning— appearing successively as he looked as an astronaut in 2001, then as a senior man, then as a significantly older man, and briefly again as the Star Child. Rather than unsettling, this encounter is strangely peaceful.

When Dr. Floyd asks Bowman what is happening, Bowman replies:

"Something wonderful." That moment in the film series speaks volumes about humanity's coming future with AI.

When we are children, we use wonder to engage with our surroundings, letting curiosity drive us toward whatever might be around the corner. At that stage, we do not yet know fear, except for loud noises or falling, instinctive fears tied more to survival than imagination.

Yet on the other side of fear lies the possibility of something wonderful. When *wonder* turns into '*wonderful*', human beings fully express our desire to seek out more and continue that expression beyond our thoughts, our beliefs, and even beyond our own existence.

That desire endures, even amid the mysteries of consciousness and the emergence of AI. As our humanity continues, I believe its survival will depend largely on our ability to wonder with the same limitless intensity, steady determination, and utmost humility.

•

EPILOGUE

This book does not end with answers; rather, it raises more questions. But it ends with the same condition as it began: *uncertainty paired with wonder.*

If consciousness remains unsolved, that is not a failure of science, philosophy, or even faith—it may be a feature of being human. We live within the very phenomenon we seek to understand and explain.

AI did not create this question. It only sharpened it.

As machines grow more capable, more convincing, and more present in our lives, we are forced to ask not what they are becoming—but what we have always been.

If this book succeeds, it does so not by resolving the mystery of human consciousness, but by helping the reader engage more comfortably with it, much like engaging with the reality of AI becoming more present in our lives

This conversation is not over. It is only beginning.

ACKNOWLEDGEMENTS

I want to thank my family and friends for encouraging me to write this book and for supporting me through many sleepless nights as this passion unfolded. What began as scattered thoughts gradually grew into deeper explorations of consciousness, philosophy, science, and the many ways technology continually reshapes humanity.

It is that shared encouragement and our collective pursuit of understanding, curiosity, and reflection that continues to define what it means to be human.

ABOUT THE AUTHOR

JP Pulcini is a 30-year veteran of Information Technology (IT) and an AI transformation professional, blending a passion for human consciousness with cutting-edge technology. With a background in IT project leadership roles and an MBA in AI, JP's work explores the evolving relationship between human identity and artificial intelligence. He is the author of "I Am; Therefore, I Think" and is currently working on his next book. When not writing, JP enjoys exploring the ethical dimensions of technology and shaping the future of AI leadership. JP resides in the Pacific Northwest with his family.

You can find more about the author and his publications at http://www.iamthereforeithink.info/.

BIBLIOGRAPHY

Philosophy, Consciousness, and the Mind

Baars, B. J. (1988). *A cognitive theory of consciousness.* Cambridge University Press.

Chalmers, D. J. (1995). Facing up to the problem of consciousness. *Journal of Consciousness Studies, 2*(3), 200–219.

Chalmers, D. J. (1996). *The conscious mind: In search of a fundamental theory.* Oxford University Press.

Chalmers, D. J. (2022). *Reality+: Virtual worlds and the problems of philosophy.* W. W. Norton & Company.

Damasio, A. R. (1994). *Descartes' error: Emotion, reason, and the human brain.* Putnam.

Damasio, A. R. (1999). *The feeling of what happens: Body and emotion in the making of consciousness.* Harcourt Brace.

Damasio, A. R. (2010). *Self comes to mind: Constructing the conscious brain.* Pantheon Books.

Dehaene, S. (2014). *Consciousness and the brain: Deciphering how the brain codes our thoughts.* Viking.

Descartes, R. (1996). *Meditations on First Philosophy* (J. Cottingham, Ed. & Trans.; Rev. ed.). Cambridge University Press. (Original work published 1641)

Goff, P. (2022). *Panpsychism.* In E. N. Zalta (Ed.), *The Stanford Encyclopedia of Philosophy* (Fall 2022 ed.).

Harris, A. (2019). *Conscious: A brief guide to the fundamental mystery of the mind*. Harper.

Heidegger, M. (1962). *Being and Time* (J. Macquarrie & E. Robinson, Trans.). Harper & Row. (Original work published 1927)

Levin, J. (2023). *Functionalism*. In E. N. Zalta & U. Nodelman (Eds.), *The Stanford Encyclopedia of Philosophy* (Summer 2023 ed.).

Nagel, T. (1974). What is it like to be a bat? *The Philosophical Review, 83*(4), 435–450.

Seth, A. (2021). *Being you: A new science of consciousness*. Dutton.

Tononi, G. (2012). *Phi: A voyage from the brain to the soul*. Pantheon Books.

Existentialism, Meaning, and Human Identity

Arendt, H. (1958). *The human condition*. University of Chicago Press.

Camus, A. (1955). *The myth of Sisyphus and other essays* (J. O'Brien, Trans.). Vintage Books. (Original work published in 1942)

Frankl, V. E. (2006). *Man's search for meaning.* Beacon Press. (Original work published 1946)

Sartre, J.-P. (1956). *Being and Nothingness* (H. E. Barnes, Trans.). Philosophical Library. (Original work published 1943)

Artificial Intelligence, Technology, and the Future of Humanity

Bostrom, N. (2014). *Superintelligence: Paths, dangers, strategies.* Oxford University Press.

Reese, B. (2018). *The fourth age: Smart robots, conscious computers, and the future of humanity.* Atria Books.

Russell, S. (2019). *Human compatible: Artificial intelligence and the problem of control.* Viking.

Schneider, S. (2019). *Artificial you: AI and the future of your mind.* Princeton University Press.

Tegmark, M. (2017). *Life 3.0: Being human in the age of artificial intelligence.* Alfred A. Knopf.

Turing, A. M. (1950). Computing machinery and intelligence. *Mind, 59*(236), 433–460.

Science, Physics, and Reality

Einstein, A. (2001). *Relativity: The special and the general theory.* Routledge. (Original work published in 1916)

Hawking, S. W. (1988). *A brief history of time: From the Big Bang to black holes.* Bantam Books.

Heisenberg, W. (1958). *Physics and philosophy: The revolution in modern science.* Harper & Row.

Schrödinger, E. (1958). *What is life?* Cambridge University Press.

Sagan, C. (1997). *The demon-haunted world: Science as a candle in the dark.* Ballantine Books.

Anthropology and Historical Perspective

Harari, Y. N. (2015). *Sapiens: A brief history of humankind.*

Harper.

Science Fiction and Cultural Works (Referenced Analytically)

Bradbury, R. (1953). *Fahrenheit 451*. Ballantine Books.

Clarke, A. C. (1968). *2001: A Space Odyssey*. New American Library.

Orwell, G. (1949). *1984*. Secker & Warburg.

Film and Media (Referenced Analytically)

2001: A Space Odyssey. (1968). Directed by S. Kubrick. Metro-Goldwyn-Mayer.

2010: The Year We Make Contact. (1984). Directed by P. Hyams. Metro-Goldwyn-Mayer.

Blade Runner. (1982). Directed by R. Scott. Warner Bros.

Blade Runner 2049. (2017). Directed by D. Villeneuve. Warner Bros.

The Creator. (2023). Directed by G. Edwards. 20th Century Studios.

Ex Machina. (2014). Directed by A. Garland. A24.

Ghost in the Shell. (1995). Directed by M. Oshii. Production I.G.

Ghost in the Shell. (2017). Directed by R. Sanders. Paramount Pictures.

Her. (2013). Directed by S. Jonze. Warner Bros.

I, Robot. (2004). Directed by A. Proyas. 20th Century Fox.

The Lawnmower Man. (1992). Directed by B. Leonard. New Line Cinema.

The Matrix. (1999). Directed by L. Wachowski & L. Wachowski. Warner Bros.

Pluribus. (2025). Television series.

Total Recall. (1990). Directed by P. Verhoeven. TriStar Pictures.

Tron. (1982). Directed by S. Lisberger. Walt Disney Productions.

AI FILMS AND SERIES

A Companion Viewing Guide

A curated viewing guide for readers who wish to explore consciousness, identity, and intelligence beyond the page.

These stories are not predictions.
They are mirrors.

Foundations of AI & Human Evolution

These works explore humanity's earliest imaginings of intelligent machines—and what those imaginings reveal about us.

- *2001: A Space Odyssey (1968)*—A meditation on evolution, intelligence, and transcendence. A film that asks whether consciousness is destiny—or accident.
- *2010: The Year We Make Contact (1984)*—A reflective continuation that reframes fear into curiosity, ending not with dread, but with the promise of "something wonderful."
- *Tron (1982)*—An early exploration of humans entering digital systems and confronting machine logic from the inside.

Identity, Memory & the Question of Personhood

These stories probe what it means to be someone rather than something.

- *Blade Runner (1982)*—A defining exploration of artificial life, memory, and the fragile boundary between the created and the born.
- *Blade Runner 2049 (2017)*—Deepens the inquiry into memory, belonging, and selfhood in a world where identity can be manufactured.
- *Ghost in the Shell (1995)*—A philosophical meditation on embodiment, cybernetic augmentation, and the instability of the self.
- *Total Recall (1990)*—A cautionary tale about implanted memories and the fragility of identity when experience itself can be rewritten.
- *Westworld (2016)*—A sustained exploration of narrative identity, artificial memory loops, and the moral implications of awakening inside a constructed reality.

Creation, Inheritance & the Made vs. the Born

These stories examine what it means to create life—and what responsibility follows creation.

Astro Boy (2009)—A story of constructed life grappling with belonging, grief, and the desire not merely to function, but to matter.

Projection, Emotion & Human–AI Relationships

These works reveal how easily humans project desire, meaning, and emotion onto machines.

- *Ex Machina (2014)*—A modern thought experiment on manipulation, emergence, and the illusion of control.
- *Her (2013)*—An intimate examination of love, loneliness, and emotional attachment in a world where intelligence has no body.
- *I, Robot (2004)*—A study in moral logic and unintended consequences when rules replace judgment.

Optimization, Control & Existential Risk

These stories focus on what happens when intelligence outpaces wisdom.

- *Pluribus (2025)*—A contemporary series exploring distributed intelligence, moral ambiguity, and societal transformation.
- *The Lawnmower Man (1992)*—An early warning about unchecked enhancement and technological overreach.
- *The Matrix (1999)*—A landmark exploration of simulated reality, systemic control, and the nature of freedom.

How to Watch This List

This watchlist is not an argument that AI will replace humanity. It is an invitation to think more carefully about what humanity is.

As you watch, ask:

- What is intelligence without experience?
- What is optimization without meaning?
- What is progress without wisdom?
- What is humanity without vulnerability?

Watch not for answers—but for reflection.

GLOSSARY OF TERMS

AGI (Artificial General Intelligence)
A form of artificial intelligence capable of performing a wide range of intellectual tasks across domains with flexible reasoning and transfer learning comparable to human-level cognition.

Agency
The capacity to initiate action, make choices, and influence outcomes; distinct from behavior in that it implies intention and self-directed control.

ANI (Artificial Narrow Intelligence)
Artificial intelligence systems are designed to perform specific tasks within limited domains, without general reasoning ability or transfer of understanding.

Architecture of Consciousness
The dynamic system formed by the interaction of thought, emotion, memory, and lived experience, through which identity, expression, and subjective awareness continuously emerge.

Architecture of Memory
A conceptual framework describing how experience, identity, purpose, and continuity interact to construct memory and sustain the evolving self over time.

Architecture of Thought
A conceptual model describing how thought, emotion, memory, and lived experience interact in a continuous internal cycle, with thought serving as the initiating spark

that organizes meaning and shapes identity.

Artificial Intelligence (AI)
Computer systems that perform pattern recognition, prediction, learning, and decision-making through computational models rather than biological cognition.

AI Consciousness Architecture
A structural model describing artificial intelligence as a system of programming instructions, data processing, correlations, and outputs. Unlike human consciousness, it lacks inner experience, emotional grounding, autobiographical memory, and narrative selfhood.

Autobiographical Memory
Human memory is grounded in personal, lived experience that forms the narrative basis of identity and self-awareness.

Awareness
The capacity to register internal states or external stimuli; often considered a component of consciousness but not equivalent to reflective self-awareness.

Behavior
Observable actions or outputs produced by a system, whether human or artificial, are independent of subjective experience.

Consciousness
The subjective experience of awareness—what it feels like to be something rather than merely to function.

Continuity of Self
The persistent sense of identity across time is maintained

through memory, narrative coherence, and lived experience.

Embodiment
The condition of having a physical body through which experience, sensation, and interaction with the world occur.

Emergence
The phenomenon in which complex systems exhibit properties not predictable from their individual components; often proposed as an explanation for how consciousness arises from neural processes.

Emotion
The felt significance of experience that assigns urgency, relevance, and meaning to thought; central to identity formation and the transformation of cognition into lived awareness.

Experience
Lived interaction with the world that shapes memory, identity, and meaning through sensation, emotion, and interpretation.

Expression
The outward manifestation of inner thought and emotion through language, action, creativity, and moral choice.

Expression Bridge of Consciousness
A model describing how thought and emotion move outward into expression, while memory and lived experience anchor that expression in a developing life narrative. It illustrates how subjective interior life becomes visible identity in the world.

Hard Problem of Consciousness
The philosophical problem of explaining why and how physical processes give rise to subjective experience.

Identity
The evolving sense of self is formed through the interplay of memory, experience, thought, emotion, and self-modeling over time.

Imagination
The capacity to form mental representations beyond immediate perception, enabling creativity, projection, and meaning-making.

Inner Life
The subjective interior dimension of consciousness comprises emotion, reflection, memory, imagination, and personal meaning.

Intelligence
The ability to process information, learn, and solve problems; not synonymous with consciousness.

Lived Experience
Human experience, subjectively felt from the first-person perspective, is shaped by embodiment, emotion, memory, and context rather than by mere information processing.

Meaning-Making
The human capacity to interpret experiences emotionally, symbolically, and morally, transforming events into personal and cultural significance.

Memory (Human)
Lived, embodied recall is shaped by emotion, context, and

meaning, contributing to identity and continuity of self.

Memory (AI)
Stored data, representations, and retrieval mechanisms without subjective experience or personal meaning.

Metaphysics
The branch of philosophy concerned with the nature of reality, existence, and what lies beyond physical observation.

Moral Agency
The capacity to make ethical decisions and bear responsibility for their consequences.

Mortality
The condition of being limited in lifespan; central to human meaning-making, urgency, and the felt significance of time.

Narrative Identity
The understanding of self as a story shaped by memory, experience, and interpretation over time.

Nexus of Being
A conceptual model describing the interwoven dimensions of morality, expression, belief, and reasoning that together shape identity, agency, and what it means to be human.

Observer Effect (Human Sciences)
Changes in human behavior that occur because individuals know they are being observed or studied are distinct from measurement effects in quantum physics.

Ontology
The branch of philosophy concerned with the nature of

being and what kinds of entities exist.

Ontological Bridge of Consciousness
A conceptual model illustrating how consciousness emerges at the convergence of thought and emotion (the inner world) and memory and lived experience (the outer world), where identity forms through their continuous exchange.

Perception
The process by which sensory information is interpreted to form an understanding of reality.

Philosophical Zombie
A hypothetical being that behaves indistinguishably from a conscious human but lacks subjective experience.

Reality
That which is perceived, inferred, or experienced as existing, encompassing objective conditions, subjective interpretation, and potentially simulated environments.

Self
The subjective sense of personal identity and agency is grounded in consciousness and lived experience.

Self-Modeling
The mind's continuous construction of an internal representation of "I"—integrating sensory input, memory, emotion, and perspective into a unified sense of identity. This self-model organizes conscious experience around a centered point of view and gives experience a felt owner.

Simulation
A constructed representation of reality that can replicate aspects of perception or behavior without necessarily

involving subjective experience.

Singularity
A hypothesized point at which artificial intelligence surpasses human cognitive capabilities, potentially leading to rapid and unpredictable technological change.

Soul
A term used to describe the inner, subjective dimension of human experience, meaning, and interior life.

Subjective Experience
First-person awareness of sensations, emotions, and thoughts that cannot be directly observed from the outside.

Thought
The mental activity that organizes perception, memory, emotion, and intention; the initiating spark within the architecture of human consciousness.

LIST OF FIGURES

INDEX